# QUANTUM THEORY OF CHEMICAL REACTIVITY

*by*

### R. DAUDEL
*Center of Applied Wave Mechanics (C.N.R.S.), Paris, France*

D. REIDEL PUBLISHING COMPANY

DORDRECHT-HOLLAND / BOSTON-U.S.A.

THÉORIE QUANTIQUE DE LA RÉACTIVITÉ CHIMIQUE

*First published in 1967 by Gauthier-Villars, Paris*

*Translated from the French by TEK Translation & International Print Ltd., London*

Library of Congress Catalog Card Number 73–75762

ISBN 90 277 0265 9

Published by D. Reidel Publishing Company,
P.O. Box 17, Dordrecht, Holland

Sold and distributed in the U.S.A., Canada, and Mexico
by D. Reidel Publishing Company, Inc.
306 Dartmouth Street, Boston,
Mass. 02116, U.S.A.

Printed in The Netherlands by D. Reidel, Dordrecht

# TABLE OF CONTENTS

# PREFACE

*Quantum Theory of Chemical Reactivity* may be read without reference to the fact that it is actually the third of three volumes of a treatise on quantum chemistry, the science resulting from the implementation of mathematical laws in the realm of molecular populations.

The first two volumes of the treatise, 'Fondement de la Chimie Théorique' and 'Structure Electrique des Molécules' were, like this third volume, originally published by Gauthier-Villars; Pergamon published the English translations of these two volumes. I am grateful to D. Reidel Publishing Company for translating the third volume of the treatise into English. Readers familiar with English rather than French now have access to the complete series.

This treatise is a reflection of the courses I taught at the Sorbonne from 1950 until 1967 to students in their second cycle (3rd and 4th year) and third cycle (5th and 6th year) working towards a doctorate in this particular field. It is based on the reading of over a thousand articles, and is intended for students as well as for physical chemists, and chemists, research workers and engineers taking an interest in quantum chemistry for its own sake or for its application in industry, pharmacology and the life sciences.

Reidel's initiative is particularly valuable because in my opinion *Quantum Theory of Chemical Reactivity* is the most important of the three volumes of the treatise. Doubtless for this reason only the third volume was published in Japanese by Baifukan, thanks to Professors Hayashi and Sohma.

The fundamental objective of quantum chemistry is the interpretation and prediction of the chemical reactivity of molecules. The present book attempts to show that it can actually be done in this important field.

# CHAPTER I

# BASIS OF THE QUANTUM THEORY OF THE CHEMICAL REACTIVITY OF MOLECULES

## 1. Introduction

### A. GENERAL POINTS REGARDING CHEMICAL REACTIVITY

Quantum chemistry is by definition the application of the methods of wave mechanics to the study of chemical phenomena, that is to say the transformation of molecules during chemical reactions. A specialist in quantum chemistry must have a very extensive knowledge of wave mechanics and must therefore also possess skill in handling the mathematical sciences, whilst at the same time he must have been trained as a chemist so as to be able to understand and interpret experimental data in the greatest possible depth. A purely theoretical scientist who is insufficiently imbued with experience in experimental chemistry often runs the risk of replacing the reaction he is studying by too simple a model and of carrying out long calculations just to reach a figure which does not represent the important element of the problem. That is why we felt it necessary to start this book with a general analysis of chemical reactivity so that we can stress the principal factors.

Essentially it is possible to draw a distinction between three modes of production of chemical reactions. In the simplest of cases the reactions are brought about by simply 'heating' the reaction mixture. By 'heating' we mean any operation which brings the reaction medium to a temperature which is higher than absolute zero. At ordinary temperatures the excited electronic levels of the molecules are not densely occupied. For this reason molecules more often than not react in their fundamental states during shocks due to thermal agitation. Consequently it is reasonable to call this type of reaction *thermochemical*.

Certain reactions are brought about as a result of the additional influence of a field. These are *electrochemical reactions*. Finally, there are others which are set off by the effect of radiations. One may talk of *radiochemical* reactions. When these radiations are photons, we are said to be dealing with *photochemical* phenomena.

### B. THE PART PLAYED BY THE WALLS

In all cases the reaction medium may be either homogeneous or heterogeneous.

So far as possible we will avoid studying reactions in a heterogeneous phase because of their greater complexity.

In most cases we will select our examples from those concerned with homogeneous gaseous or liquid media.

It should, however, be noticed that even in this case it is not possible in practice to rule out a certain degree of heterogeneity. The gas or the liquid has to be held in a vessel and *the walls of this vessel sometimes play an essential part* in the phenomenon under study.

The part so played may be demonstrated either by varying the nature of the wall or by modifying the ratio between the surface area and the volume of the container. Moreover, these two tests are necessary if one wishes to make absolutely certain that the walls do not intervene directly in a reaction.

The speeds of the thermal decomposition of dimethyl ether [1] and vinyl ether [2] thus are highly dependent on the surface area: volume ratio.

The speed of oxidation of hydrocarbons in the presence of hydrobromic acid varies a great deal according to the nature of the vessel containing the reaction medium [3]. Regarding the influence of the walls one may consult, for instance, N. N. Semenov, *Some Problems of Chemical Kinetics and Reactivity*, Vol. 1, Pergamon Press, 1958, p. 183.

### C. THE PART PLAYED BY IMPURITIES

We must also never forget that however great the care that has been taken in preparing the substances studied, they still contain *impurities* and again *the part played by these may be of paramount importance.*

Thus traces of iron or copper will catalyse the oxidation of ascorbic acid by oxygen. In order to get rid of impurities of this kind, Weissberg *et al.* [4] found it necessary to add a little cyanide and potassium thiocyanate to the reaction medium.

### D. PART PLAYED BY INERT GASES AND SOLVENTS

Finally, it would be wrong to underestimate the possible part played by substances which *a priori* do not appear to take part in the reactions, even if they are regarded as *inert*.

It is, in fact, known that the addition of a small quantity of inert gas (argon, for example) may appreciably modify the kinetics of a gas phase reaction by playing the part of a third substance during the recombination of radicals, by hindering the diffusion of the substances towards the walls or, conversely, slowing down the diffusion of the elements of a chain which has started on a wall. In the liquid phase *the solvent should very often be regarded as a true reagent*: the mechanism of a reaction may depend on the nature of the solvent and very frequently a reaction may be brought about rapidly in a certain solvent and not at all in another.

Thus Normant succeeded in preparing in tetrahydrofuran mixed organo-magnesium compounds which are not formed in ether.

### E. ANALYTICAL BALANCE AND MECHANISM OF REACTION

All these facts go to show that there are generally few relationships between the mechanism of a reaction and its analytical balance.

Thus it is well known that when we write

$$H_2 + Cl_2 \rightarrow 2HCl$$

for the photochemical reaction between chlorine and hydrogen, we do not make any claims to describing the mechanism of this reaction, the process of which is much better suggested by writing

$$Cl_2 + h\nu \rightarrow Cl + Cl$$
$$Cl + H_2 \rightarrow HCl + H$$
$$H + Cl_2 \rightarrow HCl + Cl.$$

In fact this chain reaction gives a mean quantum yield of the order of one hundred thousand. This means that a single photon gives rise on an average to the formation of one hundred thousand molecules of hydrochloric acid. The mean length of chain initiated by a single photon is therefore about one hundred thousand links.

Moreover this is a case where the part played by the wall is important since it facilitates the re-combination of the 'active centres', i.e. of the H and Cl atoms by absorbing energy and the quantity of movement

$$Cl + Cl + wall \rightarrow Cl_2 + wall$$
$$H + H + wall \rightarrow H_2 + wall.$$

This means that in this case the wall contributes to the cutting of the chains. The same is true of certain impurities like oxygen, which may cut a chain by dint of the reaction [5]

$$Cl + O_2 \rightarrow ClO_2.$$

The auto-oxidation of sulphites in aqueous solution will provide another example of a reaction where the mechanism is very different from the reaction suggested by the simple balance. This reaction is catalysed by traces of copper and slowed down by traces of alcohol.

The following mechanism has been proposed for this reaction [6–10]

$$
\left.
\begin{array}{l}
HSO_3^- + Cu^{++} \quad \rightarrow Cu^+ + HSO_3 \\
O_2 + H_2O + Cu^+ \quad \rightarrow Cu^{++} + OH^- + HO_2
\end{array}
\right\} \text{ initiation of chain}
$$

$$
\left.
\begin{array}{l}
O_2 + H_2O + HSO_3 \rightarrow H_2SO_4 + HO_2 \\
HO_2 + HSO_3^- \rightarrow HSO_4^- + OH \\
OH + HSO_3^- \rightarrow HSO_3 + OH^-
\end{array}
\right\} 
\begin{array}{l}
\text{propagation} \\
\text{of chain}
\end{array}
$$

$$
\left.
\begin{array}{l}
2HSO_3 \quad \rightarrow S_2O_6H_2 \\
2HO_2 \quad \rightarrow H_2O + O_2 + O \\
HO_2 + ROH \rightarrow RO_2H + H_2O
\end{array}
\right\} \text{ cutting of chain.}
$$

It will be observed that in the examples given above the different stages of the re-actions are never produced by the impact of a neutral molecule with an even number of electrons against another molecule with an even number of electrons. At least one of the participants in an effective collision is always either a free radical (such as H, Cl, $O_2H$, OH, $SO_3H$) or an ion (such as $Cu^{++}$). This fact holds good quite generally. *A reaction stage seldom results from the impact of a number of neutral and non-radical molecules.*

In the *homogeneous* gaseous phase the stages of the reaction may be classified into three main groups according to the number of participants involved in this particular stage.

Monomolecular stages are rare and frequently correspond to fairly complex molecules. More often than not they correspond to an isomerisation or a dissociation *

$$ABC \rightarrow ACB$$
$$ABC \rightarrow A + BC.$$

Bimolecular stages constitute by far the most frequently occurring processes of reaction.

These may be accompanied by associations, substitutions, dissociation etc.

$$AB + CD \rightarrow ABCD$$
$$AB + CD \rightarrow AC + BD$$
$$AB + CD \rightarrow A + D + BC.$$

Thermomolecular stages are less common: the most frequent of them correspond to recombinations of atoms or free radicals in the presence of a third reactant which is capable of absorbing a part of the energy and of the quantity of movement:

$$2A + BC \rightarrow A_2 + BC.$$

In the liquid phase the molecules 'roll' over one another and it is very difficult to define the notion of *molecularity*.

In the best of cases the methods of kinetics enable one to establish the *order* of a stage. In the very frequent case where the reaction takes place 'in solution' the concentration of the solvent is not generally varied. The total order of the stage envisaged therefore provides a lower limit for its molecularity as a fairly ill-defined number of solvent molecules may intervene during the collisions and increase this molecularity. There are also many cases where solvent molecules are bound to the entities taking part in the collision by precise forces of greater or lesser magnitude. This is the case with solvated ions. Sometimes the solvent even seems to form an integral part of the reacting molecules. One has every reason to regard this as being the case with mixed organo-magnesium compounds which are represented conventionally as R–Mg–X

---

* By way of example of a monomolecular process we will mention the first stage of the thermal decomposition of nitryl chloride:

$$NO_2Cl \rightarrow NO_2 + Cl$$
$$NO_2Cl + Cl \rightarrow NO_2 + Cl_2.$$

(H. F. Cordes and H. S. Johnston, *J. Am. Chem. Soc.* **76** (1954) 4264; M. Volpe and H. S. Johnston, *J. Am. Chem. Soc.* **78** (1956) 3903; N. B. Slater, *Theory of Unimolecular Reactions*, Cornell University Press, 1959, p. 175.

and which in actual fact doubtless contain molecules such as:

$$
\begin{array}{c}
\text{OEt}_2 \\
R \quad\quad X \mid \quad \text{OEt}_2 \\
\diagdown \diagup \diagdown \diagup \\
\text{Mg} \quad \text{Mg} \\
\diagup \diagdown \diagup \diagdown \\
R \quad\quad X \mid \quad \text{OEt}_2 \\
\text{OEt}_2
\end{array}
$$

in the frequent case where the ether represented here by $OEt_2$ serves as solvent [11].

F. LEVELS OF VIBRATION, ROTATION AND CONFORMATIONS

We think it would serve a useful purpose if even in this introduction we were to demonstrate a final aspect of the complexity of chemical reactions. At ordinary temperatures the molecules of one and the same type are not all in the same energy state. Most of them are usually in their fundamental electronic state, as we have mentioned, but if we are dealing with a gaseous phase it is necessary to take into consideration numerous levels of rotations and vibrations and there exists a mean statistical distribution of the molecules between these different levels accurately reflecting the temperature. In the case of liquids one must substitute for levels of rotation those less clearly defined levels of disturbed rotation known as 'libration'.

It is also as well to recall that a molecule very frequently possesses a very large number of conformational isomers which are more often than not derived from the fact that along a single bond the rotation is not entirely free, as used to be supposed, but there exist privileged positions which give rise to the same number of rotation isomers. The majority of these conformational isomers exist side by side in the reaction media. In such a way that the introduction of a molecule of a given type, say

into a reaction medium causes the appearance of a number of conformational isomers, each of them present in various states of rotation/vibration or libration/vibration. Consequently under the simple notation AB there is hidden a double *molecular population*: an isomer population and a level population.

The multiplicity of the conformations obviously frequently increases with the size of the molecules and in the case of macromolecules this aspect of the problem is an essential one and one can see from this that the reactivity of one conformation may be very different from that of another, such as for example the convolutions of the molecule may either hide or lay bare one or other particularly reactive centre. From this one may conclude that the reactional efficacy of the impact between two macromole-cules will depend a great deal on their respective conformations.

In conclusion, we have already noted the *possible importance of the walls, the part played by impurities and the intervention of the solvent during reactions where* it is not two or three molecules which impinge on one another, but *several double populations of molecules.*

### G. THE ROLE OF QUANTUM CHEMISTRY

In order to conclude this brief introduction, it remains to us to explain how quan-tum chemistry may contribute to advancing our knowledge of chemical reacti-vity.

The search for the mechanism of a reaction is carried out by successive approxima-tion. One tries first of all to determine the principal stages of it. Frequently intuition, guided by the results of previous studies of similar cases, and imagination are the essential motive elements which lead one towards the concept of mechanisms which are *a priori* compatible in the general framework of the data belonging to one's ex-perience. Usually many of the mechanisms which are presented to the mind in this way resist any qualitative criticism based on the first facts to be established. One then imagines new experiments which make use of the most elaborate methods of modern techniques (isotope effect, measurement of the specific rotation, method involving radio-indicators etc.). The new results then frequently make it possible to reduce the choice between the possible mechanisms, sometimes to the point where only one seems to be feasible. However, it is only with difficulty that one can reach a stage of cer-tainty when it comes to the question of mechanism. It is possible to demonstrate that a mechanism is wrong: but it is not possible to prove that a mechanism is correct. One may regard oneself as satisfied when one has found a mechanism which is as simple as possible and yet interprets all the known facts.

When the feasible stages have been defined in this way, what is left to do is to find out what happens during each stage, to understand what happens during each impact, how the nuclei, how the electronic density are re-organised, why a given stage of a re-action is more rapid than the corresponding stage of a similar reaction. This is where quantum chemistry intervenes for the most part. The aim of quantum chemistry is in fact first of all the study of the absolute speeds of reactions, that is to say the determina-tion of specific speed constants of the different stages, showing the part played by various factors which determine these speeds.

In practice one can only carry out an accurate *a priori* calculation of the speeds of reactions in very simple cases. But the calculation of the relative speeds of two similar reactions may give excellent results. Thus theoretical calculations give (to within a factor of two) the relative speeds of the methylation of alternant hydrocarbons [12] under certain experimental conditions. One may therefore make useful predictions from this in regard to hydrocarbons which have not yet formed the object of experi-mental studies and in particular one may predict the positions of the most reactive atoms in these molecules.

In favourable cases of this kind, quantum chemistry can supply a quantitative process for providing information regarding or confirming a mechanism.

It also makes it possible to establish the relationships between the reactivity and the structure and thus makes it possible to predict the modification of structure to be made to a molecule in order to change its reactivity in a given direction. Finally, the general laws thus established may take their place in the cluster of intuitions which lead one towards the discovery of a mechanism.

Essentially there are two major theories of reaction speeds. The first, based on the theory of collisions, is perhaps the more fundamental. But the second, which is based on the hypothesis of the state of transition, lends itself better to applications of wave mechanics. We will therefore be mainly examining this latter theory.

This theory assumes that the intermediate complex which molecules form when in a state of collision passes through a state known as a state of transition in reversible thermodynamic equilibrium with the initial products of the reaction. We shall therefore be led right from the beginning of this work to present a few results relating to reversible equilibria.

## 2. Equilibrium Constants in the Gaseous Phase

A. BASIC FORMULA

Let us consider a reversible equilibrium of the simplest type, namely

$$A \overset{K}{\rightleftharpoons} B.$$

Numerous tautomeric equilibria, for example the keto-enolic equilibrium

$$R-CH_2-\underset{\underset{O}{\|}}{C}-R' \rightleftharpoons R-CH=\underset{\underset{OH}{|}}{C}-R'$$

are of this type.

If we suppose that the law of mass action applies rigorously, we will get

$$K = \frac{[B]}{[A]}$$

if the square brackets show the concentrations of the corresponding molecules.

As we have already said, each type of molecule will show itself in the form of a double population.

Let us use $\varepsilon_{iA}$ to denote the possible energy levels of the molecules A, taking into account the conformations, rotation, vibration etc.

Boltzmann's law [13] teaches us that the number $N_{iA}$ of molecules possessing the energy $\varepsilon_{iA}$ is proportional to

$$p_{iA}e^{-(\varepsilon_{iA}/\chi T)}$$

if $\chi$ represents Boltzmann's constant and $p_{iA}$ represents the *a priori probability* of the energy state $\varepsilon_{iA}$, that is to say the number of physically different states possessing the energy under consideration:

The total number $N_A$ of the molecules A may therefore be written

$$N_A = \sum_i N_{iA} = b \sum_i p_i e^{-(\varepsilon_{iA}/\chi T)}$$

in which $b$ is the proportionality constant.

If these molecules occupy a volume $V$, one gets

$$[A] = \frac{N_A}{V} = \frac{b}{V} \sum_i p_i e^{-(\varepsilon_{iA}/\chi T)}$$

whence, by applying the same formula to the molecules B

$$K = \frac{\sum_i p_{iB} e^{-(\varepsilon_{iB}/\chi T)}}{\sum_j p_{jA} e^{-(\varepsilon_{jA}/\chi T)}} .$$

If $\varepsilon_{0A}$ and $\varepsilon_{0B}$ represent respectively the lowest possible energies for the species A and B, one may write

$$K = \frac{\sum_i p_{iB} e^{-((\varepsilon_{iB} - \varepsilon_{0B})/\chi T)}}{\sum_i p_{iA} e^{-((\varepsilon_{jA} - \varepsilon_{0A})/\chi T)}} \, e^{-((\varepsilon_{0B} - \varepsilon_{0A})/\chi T)}$$

or again

$$\boxed{K = \frac{f_B}{f_A} e^{-(\Delta\varepsilon/\chi T)}}$$

assuming

$$f_M = \sum_i p_{iM} e^{-((\varepsilon_{iM} - \varepsilon_{0M})/\chi T)}$$

$$\Delta\varepsilon = \varepsilon_{0B} - \varepsilon_{0A} .$$

The $f_M$ functions thus introduced bear the name of *partition functions*. The value of writing the formula in this way is to show the difference $\Delta\varepsilon$ between the fundamental energies of species A and B.

The role of the partition functions may be emphasized by the study of a very simple example: the equilibrium between ortho and para hydrogen.

## B. EXAMPLE OF PARA AND ORTHO HYDROGEN

The nuclei of the light hydrogen molecule $H_2$ being semi-whole spin protons constitute a system of two fermions which are capable of being either in the singlet state ($S = 0$) or in the triplet state ($S = 1$). The term para hydrogen is used to denote the molecules whose nuclei are in the singlet state and ortho hydrogen those which correspond to the triplet state.

Between these two types of molecule a thermodynamic equilibrium is established

$$H^2 \underset{\text{para}}{\overset{}{\leftrightharpoons}} \underset{\text{ortho}}{\overset{K}{}} H^2 .$$

The interaction between the spin of the nuclei and the electrons is very weak. So that as a first approximation the energy levels of the two species of hydrogen are identical.

One may write

$$\varepsilon_{iB} = \varepsilon_{iA} .$$

We therefore have

$$\varepsilon_{0B} = \varepsilon_{0A} \quad \text{whence} \quad \Delta\varepsilon = 0 .$$

But this does not mean the identity of the partition functions. The projection of the spin of the nuclei of para hydrogen on the axis of $z$ is necessarily zero. That of ortho hydrogen, on the other hand, may measure 0, $-(h/2\pi)$ or $+(h/2\pi)$. This multiplicity therefore introduces a factor three in the *a priori* probability of the states of ortho hydrogen. We therefore have

$$p_{iB} = 3p_{iA} .$$

In conjunction with the relationship already pointed out $(\varepsilon_{iA} = \varepsilon_{iB})$ this relationship makes it possible to see immediately that

$$f_B = 3f_A$$

whence:

$$K = 3 .$$

*Light molecular hydrogen in thermodynamic equilibrium therefore contains three times more ortho hydrogen than para hydrogen.*

C. EXPRESSION OF THE MACROSCOPIC THERMODYNAMIC MAGNITUDES AS A FUNCTION
   OF THE STATISTICAL THERMODYNAMIC MAGNITUDES

It is well known that there exists between the variation in standard free energy $\Delta F$ associated with an equilibrium:

$$A + B + C \rightleftharpoons G + H$$

and the equilibrium constant $K$ the relationship:

$$\Delta F = - RTLK . \tag{1}$$

Furthermore, the relationship which we have worked out

$$K = \frac{f_B}{f_A} e^{-(\Delta\varepsilon/\chi T)}$$

can easily be generalised and would, for example, be written in the case of the above equilibrium

$$K = \frac{f_G f_H}{f_A f_B f_C} e^{-(\Delta\varepsilon/\chi T)}$$

in which

$$\Delta\varepsilon = (\varepsilon_{0G} + \varepsilon_{0H}) - (\varepsilon_{0A} + \varepsilon_{0B} + \varepsilon_{0C}).$$

The ratio of the partition functions thus appearing will be noted as $f$ for reasons of simplicity.

We will therefore write

$$K = f e^{-(\Delta\varepsilon/\chi T)}. \tag{2}$$

Bringing together Formulae (1) and (2) we get

$$\boxed{\Delta F = - RTLf + \mathcal{N}\Delta\varepsilon} \tag{3}$$

if $\mathcal{N}$ represents Avogadro's number (because, as is known, $R = \mathcal{N}\chi$).

The enthalpy variation $\Delta H$ is itself connected with the constant $K$ by the formula

$$\Delta H = RT^2 \frac{\mathrm{d}LK}{\mathrm{d}T}$$

whence, taking into account (2)

$$\boxed{\Delta H = RT^2 \frac{\mathrm{d}Lf}{\mathrm{d}T} + \mathcal{N}\Delta\varepsilon} \tag{4}$$

and since the variation in entropy $\Delta S$ is connected to $\Delta H$ and $\Delta F$ by the formula

$$\Delta S = \frac{\Delta H - \Delta F}{T}$$

we can deduce from this

$$\boxed{\Delta S = RLf + RT \frac{\mathrm{d}Lf}{\mathrm{d}T}.} \tag{5}$$

The macroscopic thermodynamic magnitudes $\Delta F$, $\Delta H$ and $\Delta S$ are thus connected with the microscopic magnitudes $f$ and $\Delta\varepsilon$ of the statistical thermodynamic.

D. REGARDING THE RELATIONSHIPS OF THE EQUILIBRIUM CONSTANTS

It is frequently of interest to compare two neighbouring equilibria with one another. Let us consider on the one hand the equilibrium

$$A \underset{K}{\rightleftharpoons} B$$

and then the equilibrium

$$A' \underset{K'}{\rightleftharpoons} B'$$

in such a way that one passes for example from A to $A'$ by introducing to the molecule A a slight perturbation, let us say a substituent in order to be more precise.

Let us suppose also that it is possible to characterise the substituent by a number, say $\sigma$ (for example its electronegativity) of which the $\varepsilon_{iA'}$ and the $\varepsilon_{iB'}$ would be functions. With obvious notations we will get

$$\Delta F = - RTLK = - RTLf + \mathcal{N}\Delta\varepsilon$$
$$\Delta F' = - RTLK' = - RTLf' + \mathcal{N}\Delta\varepsilon'.$$

The quantity $\Delta F' - \Delta F$ then depends on $\sigma$ and if one develops it in a Maclaurin series we will get

$$\Delta F' - \Delta F = \left(\frac{\partial \Delta F}{\partial \sigma}\right)_{\sigma=0} \sigma + 1/2 \left(\frac{\partial^2 \Delta F}{\partial \sigma^2}\right)_{\sigma=0} \sigma^2 + \cdots$$

if by convention one inserts $\sigma = 0$ for the molecule A.

If the perturbation is slight one may content oneself with the first term and we get

$$\Delta F' - \Delta F = RTL \frac{K}{K'} = \left(\frac{\partial \Delta F}{\partial \sigma}\right)_{\sigma=0} \sigma \ldots \tag{6}$$

in which $\sigma$ is a magnitude which depends on the substituent and not on the reaction and in which

$$\left(\frac{\partial \Delta E}{\partial \sigma}\right)_{\sigma=0}$$

on the other hand represents a magnitude depending on the reaction but not on the substituent.

Equation (6) provides the basis of the method of Hammett's constants of which we shall be talking in Section 4, because it has been used mainly in a liquid medium.

Also it is frequently useful to compare the thermodynamic magnitudes of two neighbouring reactions with one another.

Let us compare in this way

$$\Delta F = - RTLf + \mathcal{N}\Delta\varepsilon$$

with

$$\Delta F' = - RTLf' + \mathcal{N}\Delta\varepsilon'$$

and

$$\Delta S = RLf + RT \frac{dLf}{dT}$$

with

$$\Delta S' = RLf' + RT \frac{dLf'}{dT}.$$

From this we get

$$\Delta F' - \Delta F = \Delta\Delta F' = - RTLf'/f + \mathcal{N}\left(\Delta\varepsilon' - \Delta\varepsilon\right)$$

$$\Delta S' - \Delta S = \Delta\Delta S = RLf'/f + RT\,\frac{dLf'/f}{dT}.$$

One therefore observed that if

$$f' = f$$
$$\Delta\Delta F' = \mathcal{N}\,\Delta\Delta\varepsilon' \tag{7}$$

and

$$\Delta\Delta S = 0. \tag{8}$$

Reciprocally, if $\Delta\Delta S$ is zero over a long temperature range, then very probably $f' = f$, whence

$$\Delta\Delta F' = \mathcal{N}\,\Delta\Delta\varepsilon'$$

the difference in the free energy variations then reflects the difference in the energy variations of a molecule during the reaction [14].

But this conclusion is no more than very probable. It cannot be applied if by chance

$$- Lf'/f = T\,\frac{dLf'/f}{dT}$$

that is to say if

$$Lf'/f = \frac{1}{dT}.$$

### 3. Rate Constants of a Bimolecular Process in the Gaseous Phase

A. THE HYPOTHESIS OF A TRANSITION STATE

In the gaseous phase the bimolecular process is the essential stage of most chemical reactions. That is why we shall start this study of the rate constants with this example.

Let us therefore assume that

$$A + B \xrightarrow{k} C + D + \cdots$$

is the process of which we are trying to assess the constant $k$. The hypothesis of the transition state [15] consists in assuming that during the course of the impact the molecules A and B merge so as to form a certain *intermediate complex*, namely $M$, a certain state of which, known as *transition state*, and called $M^{\neq}$, possesses a sufficiently long life to be in thermodynamic equilibrium with the initial products.

One will therefore write*

$$A + B \underset{}{\overset{K^{\neq}}{\rightleftharpoons}} M^{\neq} \xrightarrow{k^{\neq}} C + D + \cdots.$$

* If the reaction is reversible it would also be necessary to take into account the converse process:
$$M^{\neq} \leftarrow C + D.$$

But since, in any case, at the beginning of the reaction the concentrations of C and D are zero this process is therefore negligible and the study which we are going to develop applies both in the case where the reaction is not reversible and at its commencement.

A case where one can easily understand that it may well be like this is when the potential representing A and B is repulsive at least at the commencement of the impact. During the course of the impact the potential energy of the system will increase, and therefore its kinetic energy will decrease and consequently the nuclei which are slowing down their movement will tend to form more 'durable' configurations.

Under these conditions, the reaction speed

$$v = k [A] [B]$$

will be written thus

$$v = k^{\neq} [M^{\neq}]$$

if $k^{\neq}$ represents the rate constant of decomposition of the intermediate complex from its transition state.

Since otherwise one may generally write*

$$K^{\neq} = \frac{[M^{\neq}]}{[A] [B]}$$

one deduces from this

$$v = k [A] [B] = k^{\neq} [M^{\neq}] = k^{\neq} K^{\neq} [A] [B]$$

whence

$$\boxed{k = k^{\neq} K^{\neq}}$$

and using for the equilibrium constant $K^{\neq}$ the expression (2) which here becomes

$$K^{\neq} = f e^{-(\Delta \varepsilon^{\neq}/\chi T)} = \frac{f_{M^{\neq}}}{f_A f_B} e^{-(\Delta \varepsilon^{\neq}/\chi T)}$$

we get

$$\boxed{k = k^{\neq} \frac{f_{M^{\neq}}}{f_A f_B} e^{-(\Delta \varepsilon^{\neq}/\chi T)} .} \qquad (9)$$

B. EXPRESSION OF THE RATE CONSTANT OF DECOMPOSITION OF THE INTERMEDIATE

COMPLEX

The evaluation of $k^{\neq}$ is a very tricky operation [16]. Let us suppose that the variation in fundamental electronic energy of the intermediate complex in relation to a parameter representing the advance on the "path of the reaction" is of the type of that shown in Figure 1. We will state later on what $x$ could be.

* This method of writing neglects the perturbation caused by the disappearance of $M^{\neq}$ according to the process

$$M^{\neq} \rightarrow C + D + \cdots .$$

Various authors (R. Fowler and E. Guggenheim, *Statistical Thermodynamics*, Cambridge Univ. Press, 1939, p. 517; B. Zwolinski and H. Eyring, *J. Am. Chem. Soc.* **69** (1947) 2702) have shown that this approximation is frequently very convenient.

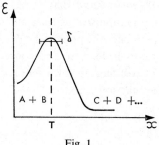

Fig. 1.

Under these conditions we will be tempted to define the transition state by the co-ordinate $x_T$ corresponding to the maximum of the potential energy.

The reaction then corresponds to the crossing of the 'potential barrier' represented by Figure 1.

The rate of reaction

$$v = k^{\neq} [M^{\neq}]$$

is then the speed at which the complex crosses the potential barrier. To evaluate $K^{\neq}$ it is consequently convenient to envisage, not only the complex in the transition state, but also the complexes which are in a neighbouring state 'crossing the potential barrier'. If then $[M^{\neq\prime}$ is the concentration of these complexes whose co-ordinates will be comprised within the interval

$$x_T - \tfrac{1}{2}\delta, \qquad x_T + \tfrac{1}{2}\delta$$

$\delta$ being a small length which is arbitrary.

Let $v^{\neq}$ be the mean speed at which the complex crosses the potential barrier, the thickness of which we have just called $\delta$. The number of complexes crossing this barrier in unit time, consequently the speed of reaction will be

$$v = k^{\neq} [M^{\neq}] = [M^{\neq\prime}] \, v^{\neq}/\delta \tag{10}$$

whence we get

$$k^{\neq} = \frac{[M^{\neq\prime}]}{[M^{\neq}]} \frac{v^{\neq}}{\delta} . \tag{11}$$

It is therefore as well first of all to evaluate the ratio $[M^{\neq\prime}/M^{\neq}]$ of the number of molecules of complexes whose co-ordinate is comprised within the interval ($x_T - \tfrac{1}{2}\delta$, $x_T + \tfrac{1}{2}\delta$) to the number of molecules of complexes in the transition state. This ratio is simply equal to the number of states "comprised within the interval $\delta$" when one takes as unity the number of states corresponding to $x = x_T$. Since the theorem of the centre of gravity applies in wave mechanics, one is led to count the number of states of a particle subjected to a constant potential in a linear space of length $\delta$, then at the top of the potential barrier the potential is more or less constant. Let therefore $m$ be the effective mass of the complex, and one knows that the energies possible for the centre of gravity will be

$$\varepsilon = \frac{n^2 h^2}{8\delta^2 m} \tag{12}$$

if we take as the origin of the energies the constant value of the potential, $n$ being an integer. One may consider these as 'translation energies' along the co-ordinate $x$.

The ratio between the number of molecules of complexes possessing this energy $\varepsilon$ and the number of molecules in the transition state will be

$$p(\varepsilon) e^{-(\varepsilon/\chi T)}.$$

Because of the smallness of the constant $h$, the energy levels will be very close together. One may therefore calculate the ratio by integrating the preceding expression instead of carrying out a discontinuous sum. One gets

$$\frac{[M^{\neq\prime}]}{[M^{\neq}]} = \int e^{-(\varepsilon/\chi T)} p(\varepsilon)\, \mathrm{d}\varepsilon,$$

$p(\varepsilon)\, \mathrm{d}\varepsilon$ representing the number of levels comprised within the interval $\mathrm{d}\varepsilon$. The number of levels being proportional to $n$, one has quite simply

$$p(\varepsilon)\, \mathrm{d}\varepsilon = \mathrm{d}n$$

from which, taking (12) into account, one finally gets

$$\frac{[M^{\neq\prime}]}{[M^{\neq}]} = \int_0^\infty e^{-(h^2/8\delta^2 m\kappa T)n^2}\, \mathrm{d}n = \tfrac{1}{2}\sqrt{\frac{\pi 8\delta^2 m\chi T}{h^2}} = \frac{\delta}{h}\sqrt{2\pi m\chi T}. \tag{13}$$

By interpolating this expression in (13) one gets

$$k^{\neq} = \frac{v^{\neq}}{h}\sqrt{2\pi m\chi T} \tag{14}$$

and we observe with satisfaction that the arbitrary length $\delta$ has disappeared. It now remains for us to calculate the $v^{\neq}$.

Let $\mathrm{d}x/\mathrm{d}t$ be the speed of a complex. Its energy $\varepsilon$ is purely kinetic (because we have taken the constant potential as the origin of the co-ordinates), we will get

$$\varepsilon = +\tfrac{1}{2}m\, \overline{(\mathrm{d}x/\mathrm{d}t)^2}.$$

One could verify this relationship directly by calculating

$$\overline{mv^2} = -\int \psi^* \frac{h^2}{4\pi^2} \frac{\partial^2}{\partial x^2} \Psi\, \mathrm{d}v$$

where the wave function $\psi$ is, as one knows, given by the expression

$$\Psi = \sqrt{2/\delta}\, \sin(n\pi x/\delta).$$

It results from this that the probability of finding for the complex a mean speed $\overline{dx/dt}$ is

$$p(\varepsilon) = e^{-1/2m(\overline{dx/dt})^2}.$$

Consequently, the mean of the speeds running in the direction of the crossing of the barrier, that is to say the speed $v^{\neq}$, if one takes all the possible states into account, is

$$v^{\neq} = \frac{\displaystyle\int_0^{\infty} \frac{\overline{dx}}{dt} e^{-1/2(m(\overline{dx/dt})^2/\chi T)}\, dn}{\displaystyle\int_{-\infty}^{+\infty} e^{-(1/2m(\overline{dx/dt})^2/\chi T)}\, dn}.$$

But since

$$\varepsilon = \frac{n^2 h^2}{8\delta^2 m} = \tfrac{1}{2}m\,(\mathrm{d}x/\mathrm{d}t)^2$$

$$(\overline{\mathrm{d}x/\mathrm{d}t}) = \frac{nh}{2\delta m}.$$

Let us assume

$$z = (\overline{\mathrm{d}x/\mathrm{d}t})$$

we get

$$\mathrm{d}n = 2\delta m/h\,\mathrm{d}z$$

whence

$$v^{\neq} = \frac{\displaystyle\int_0^{\infty} e^{-(1/2mz^2/\chi T)} z\, \mathrm{d}z}{\displaystyle\int_{-\infty}^{+\infty} e^{(1/2mz^2/\chi T)}\, \mathrm{d}z} = \sqrt{\frac{\chi T}{2m\pi}}.$$

By interpolating this result in Equation (14) we get

$$\boxed{k^{\neq} = \chi T/h.} \tag{15}$$

*Thus, when the hypotheses accepted here are satisfied, the constant of the speed of decomposition of the complex is independent of its nature.*

The expression (9) finally becomes

$$k = \frac{\chi T}{h} \frac{f_{M^{\neq}}}{f_A f_B} e^{-(\Delta \varepsilon^{\neq}/\chi T)} \tag{16}$$

and, as we shall be able to show better with an example, *all the elements shown in this formula can in principle be calculated by means of methods of wave mechanics and statistical mechanics.*

It remains for us to examine the principal cases in which the hypotheses which have helped us to establish the Formula (16) are not valid. These cases are numerous and widely different.

Let us recall first of all that we have supposed that each time the complex crosses the barrier in the left → right direction it leads to the formation of the final products.

It may, in fact, happen that the situation is more complex and that there is a probability not zero for the representative point of the system to come backwards. One then says that there is a reflection at the top of the potential barrier. One designates by the name *transmission coefficient* $\eta$ the ratio of the number of effective passages to the total number of passages in the left → right direction. The Formula (16) must therefore be corrected and now becomes

$$k = \eta \, \frac{\chi T}{h} \, \frac{f_{M^{\neq}}}{f_A f_B} \, e^{-(\Delta \varepsilon^{\neq}/\chi T)}. \tag{17}$$

As we shall see from examples, the coefficient $\eta$ is frequently of the order of unity, but it may also take values as low as $10^{-12}$. *This factor may therefore become one of the elements which determine a reaction speed.*

Another hypothesis which has served as a basis for our discussions is contained in Figure 1. We have assumed that along the reaction path the electronic energy passes through a maximum. It is easy to find examples where this is not the case. Let us consider the reaction

$$H + H \; \rightarrow \; H_2.$$

Let us study the case of the central impact and let us take as parameter $x$ the distance which separates the nuclei. We know well the form of the curve representing the variation of the electronic energy plotted against the distance $x$.

Figure 7 of our *Structure électronique des molecules** (p. 24) corresponds for example to the excellent calculation of James and Coolidge. Figure 2 (in solid lines) shows

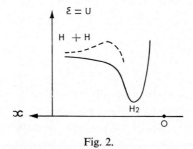

Fig. 2.

* Published in this same collection in 1962 and which we will now call the second book, the term the first book being reserved for the work *Les Fondements de la Chimie Théorique* also belonging to the same collection (1956).

the appearance of this curve represented 'upside down' since at the beginning of the reaction the parameter $x$ is very great and decreases as the reaction progresses.

This curve does not possess any maximum and the idea of transition state disappears completely in this case. Happily for the method of the transition state the probability of a central impact is low. Generally speaking the atoms will approach one another without the directions of their 'speeds' coinciding. There will therefore exist a moment of rotation of one of the nuclei in relation to the other, causing the appearance of a supplementary energy

$$\varepsilon_K = K(K+1)\frac{h^2}{8\pi^2 I}$$

in which

$$I = \mu x^2$$

(first book, p. 178).

Since $I$ increases with $x$, the energy $U_K$ is a decreasing function of $x$. By adding it to $\varepsilon$ one obtains a magnitude which varies as shown by the curve in broken lines in Figure 2.

In this way one can see the appearance of a maximum and one may consider that this corresponds to the transition state. In any case we shall study in a detailed manner the combination of hydrogen atoms (Subsection 3D).

Up to now we have assumed that the reaction can only take place if the intermediate complex crosses the potential barrier. This is a classic method of reasoning. Wave mechanics have taught us that potential barriers have a certain transparency and that therefore the reaction may also be produced by *passage underneath the potential barrier*. One then talks of a *tunnel effect*. To take this effect into account one has been led [17–19] to introduce into the Formula (17) the new factor

$$1 + t = 1 - \tfrac{1}{24}(h v_1/kT)^2$$

in which $v_1$ is a certain imaginary frequency of vibration along the axis of the values of $x$, so that $t$ is positive.

A detailed analysis of the problem has shown that $t$ is generally very small, except when the temperature is very low or when the barrier is very pointed. We therefore now write

$$k = (1 + t)\,\eta\,\frac{\chi T}{h}\frac{f_{M^{\neq}}}{f_A f_B}\,e^{-(\Delta \varepsilon^{\neq}/\chi T)}. \tag{18}$$

A last possible complication which we will mention here, without however analysing it in detail at the present moment, is bound up with the fact that even if the electronic levels of the molecules entering into the reaction are sufficiently spaced, so that only the fundamental electronic state is peopled at the temperature of the experiment*, the electronic levels of the transition state may be sufficiently close together for several

---

* And we have said that this is frequently the case.

of these levels to participate in the reaction. One then talks of a *non-adiabatic reaction* [20, 21].

## C. REACTION BETWEEN A HYDROGEN ATOM AND A HYDROGEN MOLECULE

The example of reactions arising from collisions between atoms and molecules of hydrogen will enable us to specify how one can operate in order to calculate effectively the different terms entering into the Expression (18).

One may consider the reaction for converting para hydrogen into ortho hydrogen

$$H + H_{2\,(para)} \rightarrow H_{2\,(ortho)} + H$$

or isotopic exchange reactions such as:

$$H + D_2 \rightarrow DH + D$$
$$H + H_2 \rightarrow DH + H$$
$$H + DH \rightarrow H_2 + D$$

in which D signifies deuterium.

### (a) *Calculation of the Height $\Delta\varepsilon^{\neq}$ of the Potential Barrier*

First of all we will examine the calculation of the height $\Delta\varepsilon^{\neq}$ of the potential barrier. We must therefore study the electronic energy of a system consisting of three electrons evolving in the field of three fixed nuclei of a charge $+e$. Let $r_1$ and $r_2$ be the respective distances of one of the nuclei, say $H_2$, from the two others $H_1$ and $H_3$ and let $\theta$ be the angle so formed.

The problem consists in calculating the function

$$U(r_1, r_2, \theta).$$

We immediately feel that we are dealing here with an extremely difficult problem. In point of fact, already in the case of a diatomic molecule such as $H_2$ it is necessary to make use of James and Coolidge's very complex function (first book, p. 127) in order to obtain a suitable approximation of the function $U$. For less simple diatomic molecules it is difficult to obtain a wave function which represents more than 50% of the bond energy (second book, Section 11). In the case of polyatomic molecules (second book, Section 15) the difficulties can only be more considerable.

That is why the problem has been overcome mainly with the help of semi-empirical methods: the best-known ones are due to Eyring and Polanyi [22] and to Sato [23]. These two methods have been compared by Weston [24] and by Eyring and Eyring [25].

The starting point is London's formula [26]

$$U_{ap} = Q - \sqrt{\tfrac{1}{2}[(\alpha - \beta)^2 + (\beta - \gamma)^2 + (\alpha - \gamma)]^2}$$

which gives (first book, p. 150) an approximation of the electronic energy of the system $H_3$ as a function of the total coulombic energy $Q$ and the exchange integrals $\alpha$, $\beta$ and $\gamma$ relating respectively to the pairs $(H_1H_2)$ $(H_2H_3)$ and $(H_1H_3)$.

In Eyring's method, the coulombic integral is written in the form

$$Q = Q_1 + Q_2 + Q_3$$

$Q_1$ corresponding to the pair $(H_1 H_2)$,
$Q_2$ corresponding to the pair $(H_2 H_3)$.
$Q_3$ corresponding to the pair $(H_1 H_3)$.

If the atoms $H_1$ and $H_2$ were alone, the electronic energy of the fundamental state expressed by Heitler and London's method neglecting the overlap would be (first book, p. 133):

$$U_{HH} = Q_1 + \alpha.$$

Sugiura's calculation (first book, p. 133) shows that if $E_H$ designates the energy of the ground state of a hydrogen atom, the ratio

$$\varrho = \frac{Q_1}{2E_H - U_{HH}}$$

does not vary much (between 0.15 and 0.10) when the distance between the nuclei varies from 1 to 4 Å. Taking this observation into account, Eyring's method consists in assuming that $\varrho$ is constant and in estimating $U_{HH}$ with the help of an empirical Morse curve (first book, page 178).

The knowledge of the function $U_{HH}$ and that of $\varrho$ thus makes it possible to calculate $Q_1$ and $\alpha$ for each value of $r_1$. The same method obviously applies to the calculation of $Q_2$, $Q_3$, $\beta$ and $\gamma$. By making use of London's formula one obtains $U_{ap}$ as a function of $r_1$, $r_2$ and $\theta$.

Eyring *et al.* have even thought out a very simple graphic process which makes it possible to assess the value of the radical of London's formula.

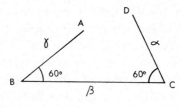

Fig. 3.

One can easily see that the distance AD measured on Figure 3 amounts to

$$\sqrt{\alpha^2 + \beta^2 + \gamma^2 - \alpha\beta - \beta\gamma - \alpha\gamma} = \sqrt{\tfrac{1}{2}[(\alpha - \beta)^2 + (\beta - \gamma)^2 + (\alpha - \gamma)^2]}.$$

For given values of $r_1$ and $r_2$ this radical will therefore be the greater, the smaller $\gamma$ is, that is to say the further away the atoms $H_1$ and $H_3$ are. The maximum of the radical will therefore occur when $H_1$, $H_2$ and $H_3$ are aligned. As this radical appears with a negative sign in London's formula, it can be shown that the *less-energy states of the complex will correspond to linear states.*

If therefore one wishes to examine 'the lowest path' of the reaction one can restrict oneself to the study of the function

$$U(r_1, r_2, \theta)$$

in which $\theta = 180°$.

Figure 4 shows the appearance of the isoenergetic lines which one obtains in that case. One can see there two potential valleys (one for a large value of $r_1$ and the other for a large value of $r_2$) separated by a sort of hill hollowed out with a little dip. The

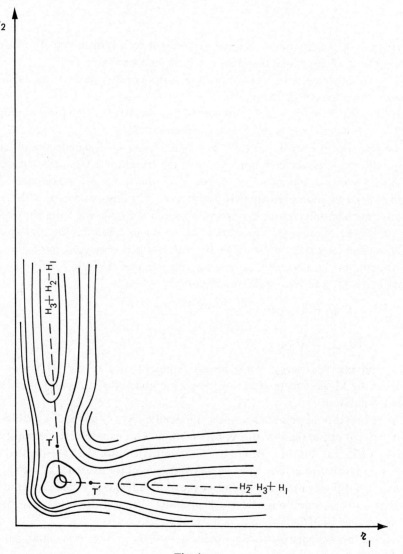

Fig. 4.

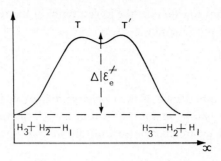

Fig. 5.

lowest path of the reaction is marked in Figure 4 by a broken line. The variation in the electronic energy along this path is represented by Figure 5.

There would therefore be two transition states corresponding to points $T$ and $T'$ situated on either side of the dip.

The bottom of the dip should correspond to a metastable state for the complex $H_3$. Such a state has not yet been observed experimentally.

One may use the term *electronic potential barrier* to designate the difference in energy $\Delta\varepsilon_e^{\neq}$ between the common energy of the transition states and that relating to the bottoms of the valleys. This energy is of the order of 14 kcal mole$^{-1}$ if one assumes $\varrho = 0.14$ and approximately 7 kcal mole$^{-1}$ if one assumes $\varrho = 0.20$.

Sato's method differs from Eyring's in two points. On the one hand Sato introduces a factor $(1+k)$ into London's formula so as to take into account the overlap integrals which we have neglected up to now. The value of $k$ is selected empirically.

Furthermore, Sato does not assume the constancy of the ratio $\varrho$. For each distance he calculates $Q_1$ and $\alpha$ from the two equations

$$U_{HH} = Q_1 + \alpha$$

and

$$U_{HH*} = Q_1 - \alpha,$$

$U_{HH*}$ signifying the energy of the lowest repulsive state of the $H_2$ system. $U_{HH}$ is estimated by Morse's method as in Eyring's method and $U_{HH*}$ is also estimated from experimental data.

The appearance of potential surfaces obtained by Sato's method resembles that already pointed out in the case of Eyring's method. One can observe two valleys separated by a hill, but this time the hill does not contain a dip. We have pointed out that there is no experimental fact which speaks in favour of the existence of such a dip. The absence of a dip may therefore be considered as an advantage of Sato's method. If one assumes $k = 0.18$, one finds $\Delta\varepsilon_e^{\neq} = 4$ kcal mole$^{-1}$ and if $k = 0.147$ one finds $\Delta\varepsilon_e^{\neq} = 7$ kcal mole$^{-1}$ [27].

We must now report the results obtained with the help of non-empirical methods [28–31].

The most important of this type is doubtless that made by Shavitt. It is based on a calculation of the function $U(r_1, r_2, \theta)$ of Boys and Shavitt. The electronic wave function of the system is constructed by the method of interaction of configurations from a base of six orbitals $1s$. In fact one attributes to each hydrogen nucleus a linear combination of two orbitals $1s$ provided with different exponents. One takes into account all the configurations which it is possible to construct on such a base, namely 34 for $r_1 = r_2$ and 60 for $r_1 \neq r_2$. One obtains

$$\Delta\varepsilon_e^{\neq} = 15.4 \text{ kcal mole}^{-1}.$$

Lippincott and Leifer (*loc. cit.*) established the potential surface from a model based on the use of Dirac's potential (second book, p. 192). They find

$$\Delta\varepsilon_e^{\neq} = 6.8 \text{ kcal mole}^{-1}.$$

Furthermore, the profile of the lowest reaction path (represented in Figure 6) causes the appearance on either side of the potential barrier of an energy minimum which would correspond to a stable $H_3$ complex.

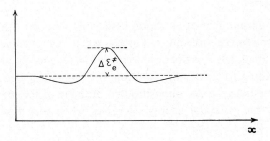

Fig. 6.

If now we wish to pass from these values of $\Delta\varepsilon_e^{\neq}$ to values of potential barrier $\Delta\varepsilon_e^{\neq}$ we are left with adding the difference between the minimum vibration energy of the transition state and the minimum vibration energy of the molecule $H_2$.

For molecules $H_2$, $D_2$, DH this vibration energy is well known and amounts to: 6.21, 4.39 and 5.38 kcal mole$^{-1}$ respectively.

It is then left for us to calculate the vibration energy of the transition state which in turn depends upon the form of the potential surface in the vicinity of the point representing this state.

In order to study the vibration movements of the complex we will use the method described in the first book (p. 180). The potential energy $U(y)$ will therefore be described as

$$U(y) = V(q_1, q_2 \ldots q_n) = V_0 + \sum_k \left(\frac{\partial V}{\partial q_k}\right)_0 q_k + \tfrac{1}{2} \sum_{k, k'} \left(\frac{\partial^2 V}{\partial q_k \, \partial q_{k'}}\right)_0 q_k q_{k'} + \cdots$$

in which the co-ordinates $q_k$ measure the displacements of the nuclei in relation to the positions which they occupy in the transition state.

We will limit the development to the quadratic terms. In all the cases under discussion so far the transition state is placed on a hill in such a way that the plane tangent to the surface of potential is horizontal. The first derivatives $(\delta V/\delta q_k)_0$ are therefore zero and if we take $V_0$ as the origin of the energies we get

$$U(y) = \tfrac{1}{2} \sum_{k,k'} \left( \frac{\partial^2 V}{\partial q_k \, \partial q_{k'}} \right) q_k q_{k'} .$$

We can therefore find normal co-ordinates $Q_1$ for which the potential energy will be written

$$U(y) = \tfrac{1}{2} \sum_l \lambda_l Q_l^2 .$$

The vibration energies are therefore given by the formula

$$\sum_l (v_l + \tfrac{1}{2}) \, h v_l$$

in which

$$v_l = \sqrt{\lambda_l}/2\pi$$

$v_l$ being the quantum number of vibration associated with the $l$-th normal mode.

Figure 7a symbolises the three normal modes which one obtains in this way for the complex $H_3$. The first mode $v_s$ is not degenerated. The second $v_l$ is also not degenerated but the corresponding frequency of vibration $v_l$ is imaginary because $\lambda_l$ is negative. $\lambda_l$ in fact corresponds to the decomposition co-ordinate of the intermediate complex and consequently the potential energy decreases when one moves away from the transition state in this direction.

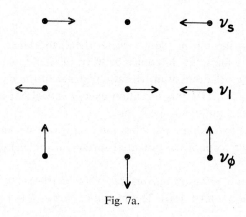

Fig. 7a.

The corresponding energy is therefore not really vibrational but translational and must therefore not be included in our calculation.

The third mode $v_\varphi$ is doubly degenerate because the deformation frequency of the complex in the two planes perpendicularly is the same.

The minimum vibration energy of the complex in the transition state may therefore be written

$$\tfrac{1}{2}hv_s + hv_\varnothing.$$

Table I then gives the values obtained for the different vibration energies (in kcal mole$^{-1}$) in the case of Eyring's method and assuming $\varrho = 0.20$.

TABLE I

|  | $\tfrac{1}{2}hv_\varnothing$ | $\tfrac{1}{2}hv_s$ | $\tfrac{1}{2}hv_s + hv_\varnothing$ | Vibration energy of the initial state | Contribution to the potential barrier |
|---|---|---|---|---|---|
| $H + H_2$ | 0.95 | 5.18 | 7.08 | 6.21 | 0.87 |
| $D + D_2$ | 0.67 | 3.66 | 5.00 | 4.39 | 0.61 |
| $H + DH$ | 0.86 | 4.57 | 6.29 | 5.38 | 0.89 |
| $D + HD$ | 0.86 | 4.38 | 6.10 | 5.38 | 0.72 |

To obtain the potential barriers it is therefore as well to add to the $\Delta\varepsilon_t^{\neq}$ already mentioned corrections of the order of a kilocalory per gram molecule. It should be noted that these corrections depend on the isotopes taking part in the reaction and therefore contribute to producing what is called the *isotopic effect*, that is to say the difference between the reaction speeds which occurs when one isotope is replaced by another.

It would now be as well to estimate the transmission coefficient $\eta$. The potential surface is seen to be symmetrical in relation to the two valleys. For symmetrical reactions of the type

$$H + H_2 \;\rightarrow\; H_2 + H$$
$$D + D_2 \;\rightarrow\; D_2 + D$$
$$D + DH \;\rightarrow\; DH + D$$

it is therefore quite normal to assume that if the complex enters the dip obtained by Eyring there are equal chances of its coming out through one or other of the two valleys. The transmission coefficient will then be

$$\eta = \tfrac{1}{2}.$$

In any case it can be seen that this coefficient here will be of the order of magnitude of unity.

The coefficient $(1+t)$ due to the tunnel effect has been assessed by Weston [32] from Sato's potential surface by means of an approximation recently proposed by Bell [33].

Table II summarises his results.

TABLE II

| $T$ | $(1+t)$ |
|---|---|
| 295 | 52.6 |
| 500 | 2.57 |
| 1000 | 1.23 |

Shavitt [34] calculated the same magnitude either using Wigner's formula [35] or using a better approximation based on Eckart's work [36].

Table III gives a few of his results. The different results of this table show that it is difficult to evaluate exactly the importance of the tunnel effect. The figures obtained depend on the method used for constructing the potential barrier and also on the formula selected for deducing the factor $(1+t)$.

TABLE III

|           | $T$  | $(1+t)$ Wigner | $(1+t)$ Eckart |
|-----------|------|----------------|----------------|
| H + H$_2$ | 200  | 5              | 180            |
|           | 300  | 2.8            | 7.6            |
|           | 1000 | 1.16           | 1.21           |
| D + D$_2$ | 200  | 3              | 13             |
|           | 300  | 1.9            | 2.7            |
|           | 1000 | 1.08           | 1.1            |

It seems, however, that this effect is far from being negligible at low temperatures and even at ordinary temperature. Bell's new formula seems to over-estimate the effect: Wigner's method, on the other hand, seems to underestimate it, particularly at low temperatures.

It will also be observed that this tunnel effect is one of the elements of the isotopic effect, a heavy isotope passing less easily under the potential barrier than a light isotope.

In order to obtain the value of the speed constant it is now left to us to calculate a ratio such as

$$f_{H_3 \neq} / f_H f_{H_2}$$

of the partition functions. In order to assess this term we have to assume that the temperature is neither too low nor too high. Below 250° for hydrogen and 150° for deuterium, certain of the approximations which we will make will cease to hold good. At temperatures which are too high the reaction could no longer be adiabatic. At ordinary temperature, for example, one may reasonably neglect for the calculation of the distribution functions all the levels where the energy exceeds by more than 0.125 eV the energy of the fundamental level. Only the fundamental electronic state of the H$_2$ molecule therefore merits consideration. The same applies to the intermediate complex.

Under these conditions the distribution function of a molecule concerned will be written

$$f_M = \sum_j p_j e^{-(\varepsilon_j / \chi T)}$$

$\varepsilon_j$ only containing the translation energies $\varepsilon_{jt}$, vibration energies $\varepsilon_{jv}$ and rotation energies $\varepsilon_{jr}$

$$\varepsilon_j = \varepsilon_{jt} + \varepsilon_{jv} + \varepsilon_{jr}.$$

One may therefore write

$$f_M = \sum_j p_{jt}e^{-(\varepsilon_{j_t}/\chi T)}p_{jv}e^{-(\varepsilon_{jv}/\chi T)}p_{jr}e^{-(\varepsilon_{jr}/\chi T)}$$

decomposing $p_j$ in accordance with the factors $p_{jt}$, $p_{jv}$ and $p_{jr}$, associated respectively with translation, vibration or rotation.

Strictly speaking the rotation energy of a molecule depends on its state of vibration, but this dependence is not very great and we can ignore it. We will therefore assume that the different types of energy are independent and we will write

$$f_M = \sum_j p_{jt}e^{-(\varepsilon_{j_t}/\chi T)} \sum_{j'} p_{j'v}e^{-(\varepsilon_{j'v}/\chi T)} \sum_{j''} p_{j''r}e^{-(\varepsilon_{j''r}/\chi T)}.$$

Let us suppose

$$f_{Mt} = \sum_j p_{jt}e^{-(\varepsilon_{j_t}/\chi T)}$$

$$f_{Mv} = \sum_{j'} p_{j'v}e^{-(\varepsilon_{j'v}/\chi T)}$$

$$f_{Mr} = \sum_{j''} p_{j''r}e^{-(\varepsilon_{j''r}/\chi T)}$$

and let us designate these quantities by the names of *distribution functions of translation, vibration* and *rotation* respectively.

The complete distribution function $f_M$ will thus appear as a product of the partial distribution functions

$$f_M = f_{Mt} \cdot f_{Mv} \cdot f_{Mr}$$

but we must never forget that this is only an approximation.

We will now learn to evaluate the different partial distribution functions. The calculation of a translation distribution function is easy.

Taking up once again a line of reasoning already proposed with regard to the crossing of the potential barrier, we will first of all study the distribution function of molecules of mass $m$ moving along a straight line of length $a$. The translation energy amounts to

$$\varepsilon_{nt} = \frac{n^2 h^2}{8a^2 m}$$

and the distribution function is

$$f_{Mt} = \sum p_{jt}e^{-(n^2h^2/8a^2m\chi T)}.$$

We have already explained why in such a case it is possible to replace the summation by an integration so that one gets

$$f_{Mt} = \int_0^\infty e^{-(n^2h^2/8a^2m\chi T)} \, dn = \tfrac{1}{2}\sqrt{\frac{8a^2m\chi T\pi}{h^2}} = \frac{\sqrt{2m\chi T\pi}}{h}\, a .$$

For molecules evolving in a volume $v$ one will therefore easily obtain

$$f_{Mt} = \frac{(2\pi m\chi T)^{3/2}}{h^3}\, v .$$

For a diatomic molecule the vibration levels have as energy

$$\varepsilon_v = hv\left(v + \tfrac{1}{2}\right)$$

in a harmonic approximation. In the distribution functions we must measure the energies in relation to that of the fundamental level, in which

$$v = 0 \qquad \varepsilon_0 = \tfrac{1}{2}hv$$

We therefore get

$$f_{Mv} = \sum_{v=0}^{v=\infty} e^{-(hvv/\chi T)} = \frac{1}{1 - e^{-(hv/\chi T)}} .$$

For a polyatomic molecule it is seen that a reasonable approximation consists in taking a product of such functions, one for each normal vibronic mode. For $H_3^{\neq}$ we therefore have to introduce

$$f_{H^{\neq}{}_{3v}} = \frac{1}{\left(1 - e^{-(hv_s/\chi T)}\right)\left(1 - e^{-(hv_\emptyset/\chi T)}\right)^2} .$$

In an analogous manner it can be demonstrated [37] that a suitable approximation for the rotation function of a linear molecule is

$$f_{Mr} = \frac{g}{\sigma}\frac{8\pi^2 I_\chi T}{h^2}$$

in which $I$ represents the moment of inertia, $g$ is the product of degenerescences due to nuclear spin and to the electronic angular moment and $\sigma$ is the number of equivalent situations which the molecule may occupy during the course of a rotation.
   Taking into account these various formulae one immediately obtains

$$\frac{f_{H^{\neq}{}_3}}{f_H f_{H_2}} = \frac{\sigma_H \sigma_{H_2}}{\sigma_{H^{\neq}{}_3}}\frac{g_{H^{\neq}{}_3}}{g_H g_{H_2}}\left(\frac{m_{H^{\neq}{}_3}}{m_H m_{H_2}}\right)^{3/2}\frac{h^3}{(2\pi\chi T)^{3/2}}\frac{I_{H^{\neq}{}_3}}{I_{H_2}} \times$$

$$\times \frac{\left(1 - e^{-(hv_{H_2}/\chi T)}\right)}{\left(1 - e^{-(hv_s/\chi T)}\right)\left(1 - e^{-(hv_\emptyset/\chi T)}\right)^2}$$

and analogous expressions for the reactions involving deuterium atoms. It will therefore clearly be seen that the ratio of the distribution functions contributes to the isotopic effect.

Introducing this expression into Formula (18) we have for the reaction

$$H + H_2 \xrightarrow{k} H_2 + H$$

$$k = (1 + t)\,\eta\, \frac{\sigma_H \sigma_{H_2}}{\sigma_{H^{\neq}{}_3}} \frac{g_{H^{\neq}{}_3}}{g_H g_{H_2}} \frac{I_{H^{\neq}{}_3}}{I_{H_2}} \frac{(m_{H^{\neq}{}_3})^{3/2}}{m_H m_{H_2}} \frac{h^2}{(2\pi)^{3/2} (\chi T)^{1/2}} \times$$

$$\times \frac{(1 - e^{-(h\nu_{H_2}/\chi T)})\, e^{-(\Delta \varepsilon^{\neq}/\chi T)}}{(1 - e^{-(h\nu_s/\chi T)})(1 - e^{-(h\nu_{\emptyset}/\chi T)})^2}$$

and analogous formulae when one or more atoms of hydrogen are replaced by a deuterium.

Now we have to compare with the experimental data the results of the various methods of the theoretical study of $k$ which we have set out. Table IV makes it possible to compare the experimental values of Geib and Harteck [38] and Farkas and Farkas [39] with those calculated on the basis of Eyring's method with $\varrho = 0.2$, ignoring the passage underneath the potential barrier.

TABLE IV

Values of $k$ in cm$^3$ mole$^{-1}$ s$^{-1}$

|  |  |  | 300 K | 1000 K |
|---|---|---|---|---|
| H + H₂ | → H₂ + H | Calculated | $7.3 \times 10^7$ | $1.5 \times 10^{12}$ |
|  |  | Observed | $9 \times 10^7$ | $2 \times 10^{12}$ |
| D + D₂ | → D₂ + D | Calculated | $3 \times 10^7$ | $0.76 \times 10^{12}$ |
|  |  | Observed |  | $1.2 \times 10^{12}$ |
| H + HD | → H₂ + D | Calculated | $2.2 \times 10^7$ | $0.52 \times 10^{12}$ |
|  |  | Observed |  | $0.68 \times 10^{12}$ |

The agreement is somewhat miraculous. The numerical coincidences must not be taken too seriously because $\varrho$ was arbitrarily fixed at 0.2. But these results do however show that with a reasonable value of $\varrho$ one may take into account both the order of magnitude of $k$ at a given temperature, so that $k$ therefore varies with the temperature, and the direction of the isotopic effect. One is therefore entitled to think that the bases of the theory are suitable: the only difficulties are to be found in the effective calculation of the magnitudes introduced by this theory.

Weston's article already quoted contains a comparison between the results of calculation and those of Sato's theory.

But since Eyring's work a certain progress has been made in the calculation of distribution functions [40] and consequently in the theoretical determination of the isotopic effects [41]. Furthermore new experimental measurements [42] have completed the 'table of values' which were known at the time of Eyring's first works. Weston has taken these various improvements into account. If one considers the set of reactions

$$H + H_2 \overset{k_1}{\to} H_2 + H$$

$$D + D_2 \overset{k_2}{\to} D_2 + D$$

$$D + H_2 \overset{k_3}{\to} HD + H$$

$$H + D_2 \overset{k_4}{\to} HD + D$$

$$H + HD \overset{k_5}{\to} H_2 + D$$

$$D + DH \overset{k_6}{\to} D_2 + H$$

table V makes it possible to make a comparison between the experimental data and the theoretical calculations* relating to the isotopic effects at 1000 K.

TABLE V

|  | $k_2/k_1$ | $k_3/k_1$ | $k_4/k_1$ | $k_5/k_1$ | $k_6/k_1$ |
|---|---|---|---|---|---|
| Experiment | 0.52 | 0.88 | 0.55 | 0.34 | 0.36 |
| Weston's calculation | 0.47 | 1.03 | 0.46 | 0.44 | 0.30 |

Once again the agreement is very suitable.

For his part, Shavitt concerned himself with a comparison of the results of his calculations with experimental results. However, he found himself in an embarrassing situation. The results of the empirical methods which conveniently reproduce the experimental results correspond to electronic potential barriers of the order of 7 kcal mole$^{-1}$ (for $\varrho = 0.2$ in Eyring's method and for $k = 0.1475$ in Sato's method).

Now Shavitt, who has no parameter since his method is non-empirical, finds

$$\Delta\varepsilon_e^{\neq} = 15.4 \text{ kcal mole}^{-1}.$$

This value is much too high to give values of the speed constants near to the experimental data. Shavitt was therefore forced to limit himself to examining the 'pre-exponential' terms of the formula giving these constants. This study does not seem to us to be sufficiently interesting to be analysed here. On the other hand, we will retain from Shavitt's discussion the disagreements which he emphasises between the different experimental measurements.

*In conclusion it appears that for the time being the use of an empirical parameter is indispensable so as to obtain an agreement between theory and experiment. But this parameter is sufficient to make it possible to take into account both the effect of the temperature and the isotopic effect.*

All this, however, permits us to assume that in the near future the non-empirical

---

* Weston chose the value 0.1475 for Sato's parameter.

calculations will also give good results. In fact it does not seem as though any important problems of principle are involved. It is probable that in the case of the reactions discussed here the theory of a transition state is suitable. It is merely a question of waiting for sufficiently powerful electronic machines so as to be able to calculate with accuracy the electronic energy of a system such as $H_3$.

D. COMBINATION OF TWO HYDROGEN ATOMS

The reaction

$$H + H \rightarrow H_2$$

on the other hand poses a trickier problem within the framework of the theory of the transition state.

We announced the discussion of this problem in Sub-section 3B and we observed that, thanks to the rotation energy of the intermediate complex, a certain potential barrier may make its appearance.

But this depends on the quantum number $K$.

One may there fore use $\Delta\varepsilon_K^{\neq}$ to denote this barrier which represents here simply the difference between the energy corresponding to the peak of the curve drawn in broken lines in Figure 2 and the energy of the two free hydrogen atoms in their fundamental state, because the complex does not possess any real vibrational frequency. The order of degenerescy of the rotation being equal to $2K+1$, the probability of realisation of the energy transition state $\Delta\varepsilon_K^{\neq}$ is proportional to

$$(2K + 1)\, e^{-(\Delta\varepsilon_K^{\neq}/\chi T)}.$$

This means that it will be necessary to replace the term $e^{-(\Delta\varepsilon^{\neq}/\chi T)}$ in Formula (18) by

$$\sum_{K=0}^{K=\infty} (2K + 1)\, e^{-(\Delta\varepsilon^{\neq}K/\chi T)}.$$

We therefore get

$$k = (1 + t)\, \eta\, \frac{\chi T}{h} \frac{f_{M^{\neq}}}{f_A f_B} \sum_{K=0}^{K=\infty} (2K + 1)\, e^{-(\Delta\varepsilon^{\neq}K/\chi T)}.$$

The distribution functions here will only contain the translation terms, the hydrogen atoms having no rotational movement nor vibrational movement and the complex having no real vibrational frequency whilst the rotation effect is already explicit.

We therefore finally get [43]

$$k = (1 + t)\, \eta\, \frac{\chi T}{h} \frac{(2\pi m_{H^{\neq}2}\chi T)^{3/2}}{(2\pi m_H \chi T)^3}\, h^3\, \frac{g_{H^{\neq}2}}{g_H g_H \sigma_2^{\neq}} \sum_{K=0}^{K=\infty} (2K + 1)\, e^{-(\Delta\varepsilon^{\neq}K/\chi T)}.$$

It is of interest to specify which are the most important terms of the sum which appears in this formula.

Figure 7b shows how

$$(2K + 1)\, e^{-(\Delta \varepsilon^{\neq} K / \chi T)}.$$

varies with $K$.

The most probable value of $K$ therefore corresponds to approximately 8. The inter-atomic distance of the transition state is therefore approximately 5 Å.

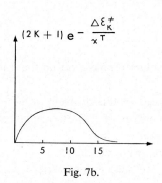

Fig. 7b.

At ordinary temperature, one finds from the calculation that

$$k = 3\,(1 + t)\, \eta\, 10^{14}\ cm^3\ mole^{-1}\ s^{-1}.$$

Furthermore, as the potential barrier is very weak, $k$ does not vary very much with the temperature.

On the other hand, this potential barrier is very wide because one passes from 5 Å to 0.74 Å between the transition state and the final state. Its transparency may there-fore be ignored at usual temperatures. We are then left with estimating $\eta$.

When two hydrogen atoms form a hydrogen molecule the electronic energy decreases by about 100 kcal mole$^{-1}$. It is obvious that whilst this energy remains in the vibronic form the molecule formed cannot be stable but has to dissociate, giving once again the initial atoms. Obtaining a stable molecule is therefore bound up with the emission of infrared radiations. The ratio between the duration of a vibration and the mean emission time therefore gives the order of magnitude of the factor $\eta$.

For 100 kcal mole$^{-1}$ the duration of a vibration is of the order of $10^{-14}$ s, the mean time of de-excitation being of the order of one second. One must therefore have

$$\eta \,\#\, 10^{-14}$$

which gives us

$$k = 3\ cm^3\ mole^{-1}\ s^{-1}.$$

It will be seen that although this reaction is effected with a very low potential barrier, it is finally much slower than the reaction $H + H_2 \rightarrow H_2 + H$ which corresponds to a much higher barrier. The slowness of the combination of two atoms is bound up with the smallness of the transmission coefficient. That is why in such cases *the termolecular processes may become more probable than bimolecular processes.*

## 4. Rate Constants of the Monomolecular and Termolecular Processes in the Gaseous Phase

### A. THE COMBINATION OF OXYGEN AND NITROGEN PEROXIDE

We have just seen that because of the smallness of the transmission coefficient the rate constant of certain bimolecular processes is very low.

In this case the termolecular processes, which are usually negligible, become on the other hand the most important ones. On the other hand, there also exist here reactions which cannot take place except with a termolecular impact. This is the case, for example, with the reactions *

$$2\,NO + X_2 \rightarrow 2\,NOX$$

in which $X_2$ may be hydrogen, oxygen or a halogen such as chlorine or bromine.

It would therefore be as well to study the behaviour of termolecular processes. In point of fact, the extension of the hypothesis of the transition state to the case of termolecular processes does not give rise to any problem. It is sufficient to assume that during the course of a triple impact an intermediate complex is formed which in the transition state is in thermodynamic equilibrium with the initial products of the reaction

$$A + B + C \rightleftharpoons M^{\neq} \rightarrow D + E + \cdots .$$

By adopting the same approximations as in Section 3 and with the equivalent notations one obtains in order to represent the rate constant of the process the expression

$$k = \eta\,(1 + t)\,\frac{\chi T}{h}\,\frac{f_{M^{\neq}}}{f_A f_B f_C}\,e^{-(\Delta\varepsilon^{\neq}/\chi T)} .$$

Let us deal [44] in greater detail with the case of one of the reactions of mineral chemistry which has been known longest

$$2NO + O_2 \xrightarrow{k} 2NO_2 .$$

The application of the methods previously described to the present reaction gives

$$k = \eta(1 + t)\,\frac{g^{\neq}}{g_i}\,\frac{\dfrac{(2\pi m^{\neq}\chi T)^{3/2}}{h^3}}{\prod^3 \dfrac{(2\pi m_i \chi T)^{3/2}}{h^3}}\,\frac{\dfrac{8\pi^2 (8\pi^3 ABC)^{1/2} (\chi T)^{3/2}}{h^3 \sigma^{\neq}}}{\prod^3 \left(\dfrac{8\pi^2 I_i \chi T}{h^2 \sigma_i}\right)} \times$$

$$\times \frac{\prod\limits_{3}^{11} (1 - e^{-(h\nu_{\neq}/\chi T)})^{-1}}{\prod^3 (1 - e^{-(h\nu_i/\chi T)})^{-1}}\,e^{-(\Delta\varepsilon^{\neq}/\chi T)} .$$

---

* As the mechanism proposed here for these reactions has been disputed, this example must be regarded mainly from the formal point of view.

In this formula the index $i$ characterises the initial states, the symbol $\neq$ designates the magnitudes associated with the transition state, A, B, C designate respectively the moments of inertia of the molecules A, B, and C. The symbols $\prod\limits^{n}$ indicate the products of $n$ distribution functions. One can easily see the nature of the different distribution functions. The intermediate complex contains six atoms and, since there are three degrees of liberty of translation, without counting the one relating to the co-ordinate associated with the path of the reaction, the corresponding distribution function contains eleven vibronic terms.

The factor which in the expression of $k$ depends on the temperature may be written

$$\frac{\prod\limits^{11}(1 - e^{-(hv^{\neq}/\chi T)})^{-1}}{T^{7/2} \prod\limits^{3}(1 - e^{-(hv_i/\chi T)})^{-1}} e^{-(\Delta\varepsilon^{\neq}/\chi T)}.$$

From this it results that

$$\frac{\mathrm{d}}{\mathrm{d}(1/T)} L \left[ k \frac{\prod\limits^{3}(1 - e^{-(hv_i/\chi T)})^{-1}}{\prod\limits^{11}(1 - e^{-(hv^{\neq}/\chi T)})^{-1}} \right] = -\frac{\Delta\varepsilon^{\neq}}{\chi I}.$$

The slope of the straight line which one must obtain by expressing the logarithm of the quantity between square brackets as a function of $1/T$ is therefore a direct measure of the potential barrier $\Delta\varepsilon^{\neq}$.

Assuming a reasonable structure for the transition state of the intermediate complex, Gershinowitz and Eyring observed that in the case of the reaction

$$2\,NO + O_2 \rightarrow 2\,NO_2$$

the potential barrier thus deduced from experimental data is approximately zero and that the constant $k$ is given approximately by the expression

$$\frac{g_{\neq}}{g_i} \prod\limits^{7}(1 - e^{-(hv^{\neq}/\chi T)}) \frac{3.2 \times 10^7}{T^3} \ (cm^3)^2 \ mole^{-2} \ s^{-1}$$

in which $\prod\limits^{7}$ represents the product related to the seven lowest frequencies of vibration of the transition state to the intermediate complex. The tunnel effect appears to be negligible and the transmission coefficient is probably in the vicinity of unity.

*It will therefore be seen that here it is the distribution functions which determine the speed of the reaction.* The appearance of the term $T^3$ in the denominator leads one to think that the rate constant runs the risk of decreasing when the temperature rises. This curious conclusion is verified by the actual calculation and also by experimental data as can be seen from Table VI.

### B. A FEW REMARKS REGARDING THE BEHAVIOUR OF THE UNIMOLECULAR PROCESSES

In Section 1 we pointed out the existence of unimolecular processes. The first stage

TABLE VI

| $T$ | $k \times 10^{-9}$ (cm$^3$) mole$^{-2}$ s$^{-1}$ | |
|---|---|---|
| | Calculated | Measured [45] |
| 80 | 86 | 41.8 |
| 300 | 3.3 | 7.1 |
| 613 | 2.1 | 2.8 |

of the thermal decomposition of nitrile chloride

$$NO_2Cl \rightarrow NO_2 + Cl$$

was given by way of example.

In point of fact, when one analyses the facts more closely one sees that a bi-molecular process practically always serves as a prelude to a unimolecular process. A molecule such as $NO_2Cl$ cannot dissociate without first of all passing into an excited state and the transfer of energy necessary is usually produced during an impact.

If one uses A* to denote the excited state involved, one will therefore usually have

$$A + A \underset{k_1}{\rightarrow} A^* + A$$

and

$$A^* \underset{k_2}{\rightarrow} B + C.$$

Corresponding to the *activation process* there is usually the inverse process

$$A^* + A \underset{k_1'}{\rightarrow} A + A$$

which is usually called the *de-activation process*.

When the activated molecule is in equilibrium with the initial products and final products of the reaction, that is to say when the speed of formation of A* is equal to its speed of disappearance, we get

$$k_1 [A]^2 = k_2 [A^*] + k_1' [A^*] [A]$$

whence we get

$$[A^*] = \frac{k_1 [A]^2}{k_2 + k_1' [A]}.$$

The speed of the reaction is then given by

$$\frac{d[B]}{dt} = k_2 [A^*] = \frac{k_1 k_2 [A]^2}{k_2 + k_1' [A]}.$$

The order of the reaction is therefore between two and one.

Obviously there are two extreme cases. If

$$k_1' [A] \gg k_2$$

$$\frac{d[B]}{dt} = \frac{k_1 k_2}{k'_1} [A]$$

the order becomes unity. This case will occur the more easily the greater $[A]$, that is to say at high pressures.

If on the other hand

$$k'_1 [A] \ll k_2$$

which always occurs at very low pressures,

$$\frac{d[B]}{dt} = k_1 [A]^2$$

the reaction is of the order two.

One must therefore expect that the *presence of a unimolecular process involves a dependence of the order of the reaction on the pressure.*

If one wishes to continue the study of the phenomenon by means of the transition state method, it is possible to identify this with the active state A* and assume that the reaction consists of a progression, starting off from this state, of the bonds which break.

One will also assume the equilibrium between the transition state and the initial products

$$A + A \underset{k'_1}{\overset{k_1}{\rightleftharpoons}} A^* + A$$

and one could suppose

$$\frac{[A^*][A]}{[A]^2} = \frac{[A^*]}{[A]} = \frac{k_1}{k'_1}$$

that is to say

$$\frac{k_1}{k'_1} = \frac{f_{A^*}}{f_A} e^{-(\Delta\varepsilon^{\neq}/\chi T)}$$

if one ignores (as in Subsection 3A) the disappearance of A* according to the process

$$A^* \underset{k_2}{\rightarrow} B + C$$

which amounts to saying that this process is slow in relation to the activation process, that is to say that the pressure remains relatively high. Therefore generally speaking one is dealing with the case where

$$\frac{d[B]}{dt} = \frac{k_1 k_2}{k'_1} [A].$$

If the reaction is written out globally in the form

$$A \overset{k}{\rightarrow} B + C$$

one has also

$$\frac{d[B]}{dt} = k[A]$$

whence

$$k = \frac{k_1}{k_1'} k_2 = k_2 \frac{f_{A^*}}{f_A} e^{-(\Delta\varepsilon_{\neq}/\chi T)}$$

and since $k_2$ is now considered as a $k^{\neq}$, one may replace it by the expression obtained in Section 3, which leads to the final form

$$k = \eta(1 + t)\frac{\chi T}{h}\frac{f_{A^*}}{f_A} e^{-(\Delta\varepsilon^{\neq}/\chi T)}.$$

Other methods have been proposed for treating unimolecular processes [46].

# References

[1] R. N. Pease, *J. Am. Chem. Soc.* **59** (1937) 425.
[2] H. A. Taylor, *J. Chem. Phys.* **4** (1936) 116.
[3] E. R. Bell *et al.*, *Ind. Eng. Chem.* **41** (1949) 2597.
[4] A. Weissberg, J. E. Lu Valle, and D. S. Thomas, *J. Am. Chem. Soc.* **65** (1943) 1934.
[5] For more details regarding the photochemical reaction of chlorine on hydrogen one should refer to G. Rollefson and M. Burton, *Photochemistry*, Prentice Hall, New York, 1939, pp. 302-312.
[6] A. Tithoff, *Z. Phys. Chem.* **45** (1903) 641.
[7] H. Backstrom and H. Alyea, *J. Am. Chem. Soc.* **51** (1929) 50.
[8] G. M. Scawab, H. S. Taylor, and R. Spence, *Catalysis*, Van Nostrand, 1937, Chapter IX.
[9] F. Haber, *Naturwissenschaften* **19** (1931) 450.
[10] J. Franck and F. Haber, *Sitzber. Preuss. Akad. Wiss. Physik. Math. Kl.* (1931) 250.
[11] See for example: Rochow, D. T. Hurd, and R. N. Lewis, *The Chemistry of Organometallic Compounds*, Wiley, 1957.
[12] R. Daudel and O. Chalvet, *J. Chim. Phys.* **53** (1956) 943.
[13] See for example: L. Boltzmann, *Leçons sur la Théorie des Gaz*, Gauthier-Villars, 1902.
[14] L. Hammett, *Physical Organic Chemistry*, McGraw-Hill, 1943.
[15] S. Glasstone, K. J. Laidler, and H. Eyring, *The Theory of Rate Processes*, McGraw-Hill, 1941.
[16] We shall present in a different manner the arguments due to H. Eyring, *J. Chem. Phys.* **3** (1935) 107; *Chem. Rev.* **17** (1935) 65; and *Trans. Far. Soc.* **34** (1938) 41.
[17] C. Eckart, *Phys. Rev.* **35** (1930) 1303.
[18] E. P. Wigner, *Z. Phys. Chem.* B**19** (1932) 903.
[19] R. P. Bell, *Proc. Roy. Soc.* **139**A (1933) 446.
[20] L. Landau, *Phys. Z. Sowjetunion* **1** (1932) 88; **2** (1932) 46.
[21] C. Zener, *Proc. Roy. Soc.* **137**A (1932) 696; **140**A (1933) 660.
[22] H. Eyring and M. Polanyi, *Z. Phys. Chem.* **12** (1930) 279.
[23] S. Sato, *J. Chem. Phys.* **23** (1955) 2465; *Bull. Chem. Soc. Japan* **28** (1955) 450.
[24] R. E. Weston, *J. Chem. Phys.* **31** (1959) 892.
[25] H. Eyring and E. M. Eyring, *Modern Chemical Kinetics*, Reinhold Publishing Corporation 1963, p. 26.
[26] F. London, *E. Electrochem.* **35** (1929) 552.
[27] R. E. Weston, *J. Chem. Phys.* **31** (1959) 892.
[28] R. Snow and H. Eyring, *J. Phys. Chem.* **61** (1957).
[29] G. E. Kimball and J. G. Trulio, *J. Chem. Phys.* **28** (1958) 493.
[30] E. R. Lippincott and A. Leifer, *J. Chem. Phys.* **28** (1958) 769.
[31] I. Shavitt, *J. Chem. Phys.* **31** (1959) 1359.
[32] R. E. Weston, *J. Chem. Phys.* **31** (1959) 892.

[33] R. P. Bell, *Trans. Far. Soc.* **55** (1959) 1.
[34] I. Shavitt, *J. Chem. Phys.* **31** (1959) 1359.
[35] E. P. Wigner, *Z. Phys. Chem.* **19** (1932) 203.
[36] C. Eckart, *Phys. Rev.* **35** (1930) 130P.
[37] See for example: S. Glasstone, K. J. Laidler, and H. Eyring, *The Theory of Rate Processes*, McGraw-Hill, 1941, p. 174.
[38] K. H. Geib and P. Harteck, *Z. Phys. Chem.* (Bodenstein Festband) (1931), 849.
[39] A. Farkas and L. Farkas, *Proc. Roy. Soc.* **152**A (1953) 124.
[40] O. L. Hershbach, H. S. Johnston, K. S. Pitzer, and R. Powell, *J. Chem. Phys.* **25** (1956) 736.
[41] J. Bigeleisen and M. G. Mayer, *J. Chem. Phys.* **15** (1947) 261.
[42] H. W. Melville and J. C. Robb, *Proc. Roy. Soc.* A **196** (1949) 445.
      M. Van Meersche, *Bull. Soc. Chim. Belge* **60** (1951) 99.
[43] H. Eyring, H. Gershinowitz, and C. E. Sun, *J. Chem. Phys.* **3** (1935) 786.
[44] According to H. Gershinowitz and H. Eyring, *J. Am. Chem. Soc.* **57** (1935) 985.
[45] E. Briner, W. Pfeiffer, and G. Malet, *J. Chim. Phys.* **21** (1924) 25.
[46] See in particular N. B. Slater, *Theory of Unimolecular Reactions*, Cornell Univ. Press, Ithaca, N. Y., 1959; G. G. Halland and R. D. Levine, in press.

# CHAPTER II

# INTERMOLECULAR FORCES, INTERACTION BETWEEN NON-BONDED ATOMS, CONFORMATIONAL EQUILIBRIA

## 1. Introduction

We have just cleared up the principal questions which form the basis of the quantum theory of chemical reactivity in the gaseous phase. We have examined this problem in the case of simple molecules. It is now a question of dealing with the reactivity of any molecules which can be very complex and extend the operation of the preceding methods to the case of liquid phases.

We have noted that generally speaking the rate constant $k$ of an elementary process is written

$$k = k^{\neq} K^{\neq}$$

in which $k^{\neq}$ signifies the constant of the rate of decomposition of the intermediate complex and in which $K^{\neq}$ signifies the equilibrium constant between the transition state of this complex and the initial products of the reaction. There are therefore two distinct problems in the calculation of a rate constant:

(a) the determination of an equilibrium constant

(b) the analysis of the rate of decomposition of a transition state.

The term $k^{\neq}$ introduces the transmission coefficient and the tunnel effect. The term $K^{\neq}$ introduces the ratio of the partition functions and the height of the potential barrier.

So as to direct ourselves towards a study of complex reactions we will have to examine what happens to these two factors in these cases. We will concern ourselves first of all with the calculation of the equilibrium constants which, even when it is not a question of the formation equilibrium of an intermediate complex, constitutes in itself a problem of chemical reactivity.

Let us therefore consider an equilibrium reaction of the type

$$A + B + C \overset{K}{\rightleftharpoons} E + F$$

for example.

We know that

$$K = \frac{f_E f_F}{f_A f_B f_C} e^{-(\Delta \varepsilon / \chi T)}$$

so long as the intermolecular interactions remain negligible.

The calculation of $K$ therefore demands the calculation of $\Delta \varepsilon$ and a knowledge of the partition functions. The calculation of the partition functions demands a know-ledge of the different energy levels (at least of the lowest) of the molecules under consideration, and therefore their different *conformations*.

The calculation of $\Delta\varepsilon$ comes down to that of the energies of the fundamental states of the different molecules under consideration. There again it will be a question of knowing the nature of the most stable conformation. *It would therefore appear that the first question which we face is the study of the conformational equilibria.*

Let us note furthermore that the methods usually employed for calculating the energy of a state of a molecule lead to fragmenting this magnitude into four parts:

(a) the energy $\varepsilon_v$ of vibration of the nuclei at absolute zero;

(b) the energy $\varepsilon_l$ of the localised bonds;

(c) the energy $\varepsilon_d$ of the delocalised bonds which usually also comprises the inter-action between the localised bonds and the delocalised bonds;

(d) the energy of interaction $\varepsilon_{n \cdot l}$ between the non-bonded atoms which contain everything which is not already included in the three first terms, and in particular the energy of steric encumberment.

We therefore have

$$\Delta\varepsilon = \Delta\varepsilon_v + \Delta\varepsilon_l + \Delta\varepsilon_d + \Delta\varepsilon_{n \cdot l}.$$

In order to terminate the generalities, all we have to do is to introduce the effect of the intermolecular actions such as, for example, the effect of the solvent. A convenient way of operating consists in assuming that, as a result of the *mean effect* of the fields created by the other molecules, the levels of the molecules studied are perturbed by a certain term $\varepsilon_s(T)$ which naturally depends on the level and the molecule under consideration. These perturbations also depend on the temperature $T$, because the thermal agitation is bound up with the temperature and acts on the mean orientation of the molecules near to a given molecule.

Under these conditions $\Delta\varepsilon$ becomes

$$\Delta\varepsilon = \Delta\varepsilon_v + \Delta\varepsilon_l + \Delta\varepsilon_d + \Delta\varepsilon_{n \cdot l} + \Delta\varepsilon_s(T)$$

and $K$ is written

$$K = \frac{f_B^s f_F^s}{f_A^s f_B^s f_C^s} \exp\left[ - \frac{\Delta\varepsilon_v + \Delta\varepsilon_l + \Delta\varepsilon_d + \Delta\varepsilon_{n \cdot l} + \Delta\varepsilon_s(T)}{\chi T} \right].$$

The indices $s$ located near the partition functions recall that the intermolecular forces modify the partition functions not only by changing the values of the levels but also sometimes by transforming the nature of the problem, as rotation functions may be converted into libration functions.

In the second book we have already examined in detail the calculation of terms such as $\varepsilon_l$ or $\varepsilon_d$ and even $\varepsilon_v$. We will have a few additions to make to this subject and we will do this as need arises.

On the other hand, we have not yet discussed the calculation of the interactions between non-bonded atoms nor intermolecular interactions.

We will therefore have to examine these questions. In any case it is found that the interaction between non-bonded atoms may be to some extent considered as a limit case of intermolecular interaction.

## 2. Brief Analysis of the Methods of Calculation of
## Intermolecular Interactions

Examination of intermolecular interactions must be envisaged from two points of view.

In the first place it is as well to examine the nature of the interactions which take place between two molecules. In principle this problem does not pose any questions because one can always treat the group of two molecules as a single molecule, a *supermolecule*. In practice, this method of attacking the problem is not very fruitful in general cases, because it leads to extremely complicated calculations and the validity of the most usual approximations needs to be discussed. The fact that the nuclei of the supermolecule evolves in vast domains, for example, makes it precarious to use Born and Oppenheimer's approximation.

It will therefore be as well briefly to examine the methods, which are frequently semi-empirical, which have been proposed for avoiding these pitfalls.

Another vantage point which it is useful to use consists in regarding the whole of a molecular population and the mean effect on a molecule of the population resulting from the presence of the others. In this case it is therefore a question of the application of statistical methods.

A. CLASSIFICATION OF INTERMOLECULAR FORCES

It is convenient to classify the intermolecular forces into a number of groups. One generally draws a distinction between the electrical forces and the magnetic forces. In practice the latter are only important when one of the partners is paramagnetic (free radical, triplet state etc.). We will not talk of this type of force here.

The electrical forces in turn are usually classified in two distinct categories: the forces with a long radius of action and the forces with a short radius of action. In this case the frontier is not very well defined. This results from the fact that although theoretically the space occupied by a molecule is infinite, the domain where there is an appreciable probability of meeting its electrons is very limited.

One may in practice, for example, fix this domain inside spheres centred on each nucleus and having as their radius the Van der Waals' radius of the corresponding atom. One will then say by convention that the forces with a short radius of action are those which come into being when the domains thus associated with the molecules have a common part, and that one is dealing with forces with a long radius of action in the opposite case.

A distinction is drawn between *three types of electrical force with a long radius of action*. The *electrostatic forces* are those which have their origin between polar molecules, that is to say are endowed with permanent moments if they were not perturbed. The *forces of polarisation* result from the fact that in reality every polar molecule creates an electrical field which polarises any neighbouring molecule; there thus appear induced moments and it is the energy of interaction between the induced moments which causes the existence of these forces.

Finally the *forces of dispersion* have a less obvious origin. They result from the fact that at a given instant even an apolar molecule generally possesses instantaneous moments which favour by correlation the presence in a neighbouring molecule of certain instantaneous moments corresponding to attractive situations. From the interaction between these moments there come into being these forces which are also known as London's forces.

A distinction is also drawn between *three types of force with a short radius of action.* The *forces of exclusion* which make themselves felt particularly when two saturated molecules approach one another (in the sense that all the places available for electrons are occupied) because of the repulsion which exists between two electrons of the same spin. The *charge transfer forces* correspond on the other hand to the case where available places exist on one of the molecules and where consequently a small fraction of the electronic density of one of the molecules slides into the domain of the other.

Finally, the *hydrogen bond* constitutes the last type of forces which are considered as intermolecular.

### B. ELECTROSTATIC FORCES

The electrostatic forces, being those which result from the coulombic interaction between two polar molecules which are supposed not to be perturbed, are in principle very easy to calculate when one knows the distributions of the charge densities $\varrho_1(M_1)$ and $\varrho_2(M_2)$ in each molecule.

The energy of interaction is obviously written

$$E_{el} = \int \frac{\varrho_1(M_1)\,\varrho_2(M_2)}{r_{12}}\, dv_1\, dv_2.$$

From the practical point of view the calculation of this integral may be laborious and when the molecules are not too close together it may be interesting to use the development on two centres of this energy.

One obtains this expression by developing the potential due to one of the distributions on a centre situated inside the other by means of Neuman's method.

One finds for example that at a point situated outside a distribution the potential $F(r, \theta, \varphi)$ due to this distribution $\varrho$ is expressed in the form

$$F(r, \theta, \varphi)\quad \sum_{n=0}^{\infty} \sum_{m=-n}^{+n} \frac{(n-|m|)!}{(n+|m|)!}\frac{Q_n^m}{r^{n+1}}\, P_n^m(\cos\theta)\, e^{-im\varphi}$$

where the $P_n^m$ are associated polynomes of Legendre and in which

$$Q_n^m = \int \varrho(r, \theta, \varphi)\, r^n\, P_n^m(\cos\theta)\, e^{im\varphi} r^2 \sin\theta\, dr\, d\theta\, d\varphi.$$

We will observe that with this last definition the charge $q$, the dipolar moment $\vec{\mu}$ and the quadrupolar moment $\overset{\Rightarrow}{Q}$ are given by the equations

$$q = Q_0^0; \qquad \mu_z = Q_1^0; \qquad \mu_x = \tfrac{1}{2}[Q_1^1 + Q_1^{-1}]; \qquad \mu_y = \tfrac{1}{2}[Q_1^1 - Q_1^{-1}]$$
$$Q_{zz} = 2Q_2^0; \qquad Q_{xx} = -Q_2^0 + \tfrac{1}{4}[Q_2^2 - Q_2^{-2}]$$
$$Q_{yy} = -Q_2^0 - \tfrac{1}{4}[Q_2^2 - Q_2^{-2}]; \qquad Q_{xz} = \tfrac{1}{2}[Q_2^1 + Q_2^{-1}]$$
$$Q_{yz} = 1/2i[Q_2^1 - Q_2^{-1}]; \qquad Q_{xy} = 1/4i[Q_2^2 - Q_2^{-2}].$$

One finally arrives at [1]

$$E_{\text{el}} = \sum_{n_a=0}^{\infty} \sum_{n_b=0}^{\infty} \sum_{m=-n<}^{n<} \frac{(-1)^{n_b+m}(n_a+n_b)!}{(n_a+|m|)!\,(n_b+|m|)!\,r_{ab}^{n_a+n_b+1}} \, Q_{na}^{m*} Q_{nb}^{m}.$$

The indices $a$ and $b$ characterise the quantities associated respectively with the two distributions $\varrho_1(M_1)$ and $\varrho_2(M_2)$. $r_{ab}$ denotes the distance of the points of development of the distributions. $n<$ indicates the smaller of the two numbers $n_a$ and $n_b$. Let us state finally that one uses axes of the colinear $z$, parallel axes of $x$ and $y$ respectively.

Thanks to this expression the calculation of the bicentric sextuple integrals is brought down to that of monocentric triple integrals. Furthermore, when the distance which separates the molecules is great, the series is rapidly convergent and one may be satisfied with the first few terms.

If one considers, for example, the case of four charges represented in Figure 8,

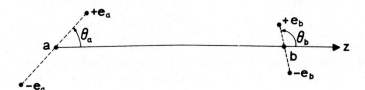

Fig. 8.

the first term of the series will be written

$$E_{\text{el}}^0 = \frac{\mu_a \mu_b}{r_{ab}^3}(-2\cos\theta_a \cos\theta_b + \sin\theta_a \sin\theta_b \cos\varphi)$$

if $\mu_a$ and $\mu_b$ signify the corresponding dipolar moments and $\varphi$ signifies the angle of the planes passing through the axis of the $z$ and containing one of the dipoles. For distances separating the dipoles at least five times greater than their lengths, this first term gives at least within $10\%$ the value of the total energy $E_{\text{el}}$.

## C. POLARISATION FORCES

To calculate the electrostatic forces we have assumed that the molecules were not perturbed. In point of fact it is not like this. Any polar molecule creates a field which polarises the other and it is known that as a first approximation under the influence of an electric field $\vec{h}$ a molecule takes an induced dipolar moment

$$\vec{\mu} = \vec{\alpha} \cdot \vec{h}$$

if $\vec{\alpha}$ signifies the polarisability tensor, the energy of the dipole induced in the field $\vec{h}$ being

$$-\tfrac{1}{2}(\vec{h} \cdot \overset{\leftarrow}{\vec{\alpha}} \cdot \vec{h}).$$

The energy of interaction due to the mutual polarisation of the two molecules may therefore be written

$$E_{\text{pol}} = -\tfrac{1}{2}\vec{h}_a \overset{\leftarrow}{\alpha}_b \vec{h}_a - \tfrac{1}{2}\vec{h}_b \overset{\leftarrow}{\alpha}_a \vec{h}_b$$

in which $\vec{h}_a$ signifies the field due to molecule 2 near molecule 1, $\vec{h}_b$ signifies the field due to molecule 1 near to molecule 2 and $\vec{\alpha}_a$ and $\vec{\alpha}_b$ signify the polarisabilities of the two molecules. In the case where the polarisability and the quadrupolar moment come down to simple scalars $\alpha_a$, $\alpha_b$, $Q_a$, $Q_b$, one arrives at the expression

$$
\begin{aligned}
E_{\text{pol}} = &-\frac{q_a \alpha_b + q_b^2 \alpha_a}{r^4} - \frac{2 q_a q_b \alpha_b \cos \theta_a + 2 q_b \mu_b \alpha_a \cos \theta_b}{r^5} \\
&-\frac{\mu_a^2 \alpha_b (3 \cos^2 \theta_a + 1) + \mu_b^2 \alpha_a (3 \cos^2 \theta_b + 1)}{2 r^6} \\
&-\frac{3 q_a Q_a \alpha_b (3 \cos^2 \theta_a - 1) + 3 q_b Q_b \alpha_a (3 \cos^2 \theta_b - 1)}{2 r^6} + \cdots .
\end{aligned}
$$

The terms in $1/r^4$ and $1/r^5$ are frequently the most important in the interaction between an ion and neutral molecule. The term in $1/r^6$, on the other hand, is essential in the interaction between a molecule and another molecule endowed with a permanent dipolar moment.

### D. DISPERSION FORCES

Let us consider to start off with the example of two hydrogen atoms. Since these do not possess permanent moments the only forces which can intervene at great distances are by definition the forces of dispersion. Let A and B be the nuclei which are supposed to be fixed and located on the axis of the $z$. Since the distance $r_{AB}$ is supposed to be great one may adopt the following unperturbed function

$$\Phi = 1s_A(1)\, 1s_B(2) + 1s_A(2)\, 1s_B(1)$$

to represent the fundamental state of the system. One may even be content with the first term of this function, because cross products such as

$$1s_A(1)\, 1s_B(2)\, 1s_A(2)\, 1s_B(1)$$

are practically nil at any point. This is in fact what we will do.

The perturbation operator which it is convenient to apply to this function is, with the usual notations

$$H' = -\frac{e^2}{r_{1B}} - \frac{e^2}{r_{2B}} + \frac{e^2}{r_{AB}} + \frac{e^2}{r_{12}}.$$

Let us assume

$$R = r_{AB}.$$

One can develop this operator in a Taylor series.

One easily finds

$$H' = \frac{e^2}{R^3}(x_1x_2 + y_1y_2 - 2z_1z_2) + \frac{e^2}{R^4}(\ldots) + \cdots$$

if $x_1, y_1, z_1$ refer to the first electron and to the nucleus 1, $x_2, y_2, z_2$ to the second electron and to the nucleus 2. Always in the case where $R$ is large one may content oneself with the first term. One then finds that the energy $E'_{disp}$ resulting from the perturbation of the first order is strictly zero. It is then as well to calculate the term of the second order. To obtain the order of magnitude of the latter one can restrict oneself to evaluating [2]

$$\frac{(H'^2)_{00}}{E_0^0}.$$

In this expression $E_0^0$ denotes the energy of the unperturbed fundamental state, that is to say in this case the energy of two atoms of hydrogen. We therefore have

$$E_0^0 = -\frac{e^2}{a_0}.$$

Furthermore it has been supposed that

$$(H'^2)_{00} = \int \Phi^* H'^2 \Phi \, dv.$$

One finds that after integration the cross terms of the operator $H'^2$ have a zero contribution such that

$$(H'^2)_{00} = \frac{e^4}{R^6} \int \Phi^* (x_1^2 x_2^2 + y_1^2 y_2^2 = 4z_1^2 z_2^2) \Phi \, dv.$$

The calculation of this integral is easy and gives

$$(H'^2)_{00} = 6 \frac{e^4}{R^6} a_0^4.$$

From this one gets

$$E''_{disp} = \frac{(H'^2)_{00}}{E_0^0} = -\frac{6e^2 a_0^5}{R^6}.$$

This expression represents to some extent an *asymptotic value* of the dispersion energy between two atoms of hydrogen when the system is *in the singlet ground state*.

Different variational treatments have been proposed by Slater and Kirkwood [3], Hassé [4], Pauling and Beach [5] and others. The best lead to a replacement of the factor 6 by the factor 6.5 in the expression of $E''_{\text{disp}}$. For his part, Margenau [6] took up again the perturbation method by taking a more complete development of the perturbation operator.

He obtained:

$$E_{\text{disp}} = - \frac{6e^2 a_0^5}{R^6} - \frac{135 e^2 a_0^7}{R^8} - \frac{1416 e^2 a_0^9}{R^{10}} + \cdots .$$

This formula gives results which agree well with experimental data.

Analogous treatments have been carried out in the case of two atoms of helium [3, 4, 6].

According to these treatments it is suitable in this case to adopt

$$E_{\text{disp}} = - \frac{1.4 e^2 a_0^5}{R^6} .$$

London [7] proposed a process for estimating the dispersion energy which is applied to atoms or even to molecules which are sufficiently near to spherical symmetry and may be different.

He obtains

$$E_{\text{disp}} = - \frac{6 n_A n_B e^4 \overline{z_A^2}\,\overline{z_B^2}}{R^6 (I_A + I_B)}$$

in which $n$ represents the number of very polarisable electrons, the $\overline{z^2}$ the mean of the square of their coordinate along the internuclear axis measured from the respective nuclei and the $I$ represent the first ionisation energies.

Since in this same type of approximation one may represent the polarisabilities $\alpha$ by the formula

$$\alpha = \frac{2 n e^2 \overline{z^2}}{I}$$

we get once again

$$E_{\text{disp}} = - \frac{3}{2} \frac{\alpha_A \alpha_B}{R^6} \frac{I_A I_B}{I_A + I_B} .$$

Applied to hydrogen, this formula gives us

$$E_{\text{disp}} = - \frac{7.6 e^2 a_0^5}{R^6}$$

(instead of $-(6.5 e^2 a_0^5 / R^6)$ in the best variational treatment).

Hornig and Hirschfelder [8] examined in the case of two non-identical systems with spherical symmetry the value of the coefficients of the terms in $1/R^8$ and $1/R^{10}$ which it is advisable to add to the term in $1/R^6$ when the distance is no longer very great.

Casimir and Polder [9] on the other hand, examined what happens at very great distances. The fact (neglected up to now) that the electromagnetic interactions are not propagated at an infinite speed, means that when the distances are great (say, from 200 Å) a decrease in what there would be without that of the interaction. This phenomenon is hardly of interest except in the case of macromolecules.

London [10] has also concerned himself with the case of two molecules each of which possesses only one axis of symmetry. By taking the mean of all the possible orientations he found that at great distances one may write

$$E_{\text{disp}} = -\frac{2}{3R^6} [A + 2(B + B') + 4C] \frac{I_A I_B}{I_A + I_B}$$

in which

$$A = \frac{\alpha_{\|A} \alpha_{\|B}}{4} \qquad B' = \frac{\alpha_{\perp A} \alpha_{\|B}}{4}$$

$$B = \frac{\alpha_{\|A} \alpha_{\perp B}}{4} \qquad C = \frac{\alpha_{\perp A} \alpha_{\perp B}}{4}$$

in which $\alpha_{\|}$ designate the polarisabilities along the axis of symmetry and the $\alpha_{\perp}$ indicate the polarisabilities measured along a direction at right angles to this axis.

Davies and Coulson [11] studied the energy of interaction due to the forces of dispersion between two conjugated molecules.

They calculated this energy by using the rigorous formula of the theory of perturbations of the second order.

According to this theory the dispersion energy is written

$$E_{\text{disp}} = \sum_{l}' \frac{H'_{0l} H'_{l0}}{E_0^0 - E_l^0}$$

in which $H'$ designates the operator representing the interaction between the two molecules and in which

$$H'_{0l} = \int \Phi_l^* H' \Phi_0 \, dv$$

$\Phi_1$ represents an excited state of the system formed by the two unperturbed molecules and $\Phi_0$ denotes the fundamental state of the same system. $E_1^0$ and $E_0^0$ represent the energies associated with these two states respectively.

The calculations have in fact been carried out in the case of the polyenes. The functions $\Phi$ have been determined in Hückel's approximation (second book, page 165). In the evaluation of the $H'_{01}$ the integrals with three and four centres have been ignored. Davies and Coulson observed that the most important term $H'_{01}$ corresponds to the case where the two molecules are in the first excited state.

It has been possible to carry out the actual calculations for different relative orientations of the two molecules of the same nature. When they are parallel to one another and perpendicular to the axis passing through their centres, the energy $E_{disp}$ varies inversely with the sixth power of the distance which separates them, namely $y$, provided that one has

$$\gamma/L > 1.8$$

($L$ designating the length of each molecule). When the values of $y/L$ are low, the energy $E_{disp}$ becomes weaker than would have been supposed from the $1/y^6$ law and is more or less proportional to $L$.

For values of $y/L$ exceeding 2 one may represent the dispersion energy approximately by the formula

$$E_{disp} = - 16 \frac{e^4 l^4}{y^6 \pi^9 \beta} L^5$$

in which $\beta$ is the usual resonance integral and $l$ is the length of a CC bond.

If, whilst keeping the molecules in parallel planes, without changing the position of their centres, one allows them to rotate in relation to one another, one finds that the maximum interaction energy corresponds to the case where the angle formed by their axes is between $60°$ and $90°$ (which is the case of hexatriene). If now one moves one of the molecules whilst maintaining it parallel to the other, one observes that the dispersion energy passes through minima and maxima. Viewed as a whole, the phenomenon is extremely complex and one may question the validity of the results which we have just described because of the approximations introduced into the calculation.

Longuet-Higgins and Salem [12] have examined the case of long saturated molecules. They have shown that in this case the interactions due to the dispersion forces are *locally additive*: this means to say that as a first approximation one may calculate the interaction energies between the different pairs of localised bonds and then add together the results obtained.

In the case of two saturated chains of the same nature and parallel to one another, with a total length of $L$, separated by a distance $D$ and formed by the repetition of a motif of length $l$, the application of the above rule leads to the expression

$$E_{disp} = \frac{A}{4l^2 D^4} \varrho \left( 3 \tan^{-1} \varrho + \frac{\varrho}{1 + \varrho^2} \right)$$

in which $\varrho$ replaces $L/D$ and in which $A$ designates a constant which depends on the structure of the motif of which the chains are built up.

In the case of saturated hydrocarbons, for example, $A$ has been estimated at approximately $- 1340$ kcal per g-molecule and per Å [13, 14].

If $D$ is very much smaller than $L$, the formula is simplified and becomes

$$E_{disp} = A \frac{3\pi}{8l} \frac{N}{D^5}$$

if $N$ denotes the number of motifs contained in a molecule.

Salem [15] used this formula to calculate the interaction energy which arises between the molecules of a fatty acid in monomolecular film. In the case of stearic acid $CH_3-(CH_2)_{16}-CO_2H$ the distance $D$ is then 4.8 Å, the dispersion energy is $-8.4$ kcal mol$^{-1}$. In the case of isostearic acid $CH_3-CH-(CH_2)_{14}-CO_2H$ the distance

$$CH_3$$

$D$ is increased to 6 Å, the dispersion energy is reduced to $-2.8$ kcal per g-molecule. One will observe first of all the importance of the interaction which is equivalent to several long hydrogen bonds. One will then note the vigour with the intensity of the phenomenon varies according to the distance $D$, because the energy is reduced by $\frac{2}{3}$ of its initial value for a simple increase of $\frac{1}{5}$ of the distance $D$. This fact enables us to understand why the forces of cohesion may be highly perturbed by a detail in structure and shows us the importance of interactions between unattached atoms on the tertiary structure and even secondary structure of macromolecules such as proteins.

To sum up, it may be stated that the forces of dispersion vary in $1/R^6$ in the simple cases, when the distance $R$ is great in relation to the size of the molecules under consideration. When the distances are very great, the fact that the electromagnetic interactions are propagated at a finite speed may involve a variation in $1/R^7$.

When the intermolecular distance is no longer very great in relation to the size of the molecules envisaged, the properties of the dispersion interactions depend to a large extent on the structure of the molecules and their relative positions. In the case of saturated molecules one observes a local additivity of the interaction energies and a dependence in $1/R^5$ on the distance which separates them.

In the case of conjugated molecules, the phenomena become very complex. The interaction increases very markedly with the size of the delocalised systems and it depends very specifically on the relative positions of the molecules in relation to one another, thus tending to create privileged arrangements.

## E. EXCLUSION FORCES

It will be seen quite clearly from Subsections 2B, C and D of this chapter that if, when the distance is great, the interactions may be represented by simple formulae (that is for example the law in $1/R^6$ for dispersion energies) which do not depend very much on the nature of the molecules present (except if they are very large), when the distance is short, the personality of the elements of the system plays an essential part. We must therefore now expect that the role of this personality is even more marked in the determination of the properties of the forces with a short radius of action which we are now going to study.

That is why before trying to set out a few general ideas it seems essential to us to examine a few examples.

The simplest of these is without doubt the example of two atoms of hydrogen. We know that for this system the lowest energy state corresponds to a singlet state $^1\Sigma$ and the second energy state corresponds to a triplet state $^3\Sigma$.

We have studied how the interaction energy of these states varies with the interatomic distance (see for example: first book, sections 21 and 22 and second book, section 6) for very low values of this distance (0 to 2 Å). In this domain the state $^1\Sigma$ comprises a minimum of energy which corresponds to a strong bond which accompanies the formation of the hydrogen molecule. The wave function which gives results which agree best with experimental data is that of James and Coolidge. We will point out in passing that since the publication of the second book a remarkable study of this question has been carried out by Kolos and Roothaan using a James and Coolidge function containing 40 terms [16].

In this same range of interatomic distance the state $^3\Sigma$ is, on the other hand completely repulsive and the energy calculated by James, Coolidge and Present [17] may be represented approximately by the interpolation formula

$$E_{p.r.} = 349.7\,(r_{AB}/a_0)^{-4/3}\,e^{-(0.357/a_0)r_{AB}}\ \text{kcal mole}^{-1}.$$

Hirschfelder and Linnett [18] studied the energy of the same states using a variational method which is particularly adapted to the calculation of dispersion energies. They think that their results are particularly suitable for distances ranging from 1.5 to 6 Å.

The curves of Figure 9 show in diagram form a synthesis of the data of James, Coolidge, and Present and those of Hirschfelder and Linnett. The curve relating to the state $^3\Sigma$ comprises an energy minimum of $-0.0074$ kcal mol$^{-1}$ for: $r_{AB}=8.45\,a_0$.

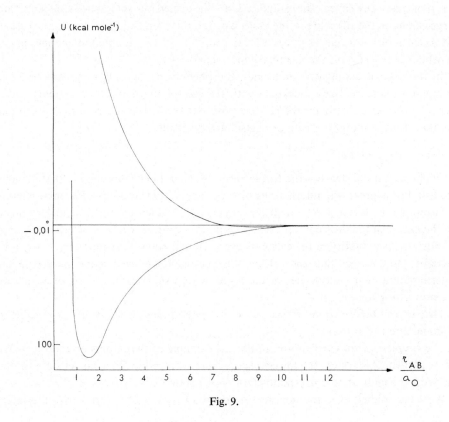

Fig. 9.

From the point of view of the electronic energy one may therefore obtain a stable configuration for this distance, corresponding to what one could call a Van der Waals association and consequently the radius $\frac{1}{2}r_{AB}=4.22\,a_0$ could be designated as the Van der Waals' radius for hydrogen atoms in the state $^3\Sigma$. It will be seen that forces with a short radius of action for this state $(r_{AB}<2\text{ Å})$ are distinctly repulsive, whilst in the same range the state $^1\Sigma$ comprises a highly attractive region. The passage from the singlet state to the triplet state therefore involves the appearance of new 'forces' of repulsion which are sometimes given the name of *Pauli's exclusion forces*.

Let us now examine the case of a system consisting of two atoms of helium. Rosen [19] calculated the interaction energy of the system $E_p$ with short distances by representing each atom of helium by means of a function of the form

$$\Psi = e^{-(2.1r_1/a_0)}e^{-(1.19r_2/a_0)}$$

and evaluating the correction made in the energy by the perturbation of the first order by introducing the exact operator. One may interpolate his results with the help of the formula

$$E_{p.r.} = 9.25 \times 10^{-10}\,e^{-4.4r_{AB}}\ \text{ergs}$$

if $r_{AB}$ is expressed in Å.

We have already pointed out that when the distances are great this energy becomes negligible and that it is necessary to take into account the perturbation of the second order which corresponds to the forces of dispersion.

Margenau [20] obtained

$$E_{disp} = -\frac{1.39 \times 10^{-12}}{r_{AB}^6}\ \text{ergs}$$

for the firm term relating to this energy.

Page [21] determined the term in $1/R^8$ and found

$$E'_{disp} = -\frac{3.0 \times 10^{-12}}{r_{AB}^8}\ \text{ergs}.$$

We will recall that the dispersion energy is usually calculated by ignoring the terms connected with the exchanges of the electronic coordinates, because the latter are negligible when the distances are great. If, however, we wish to construct a complete curve representing the variation in the interatomic energy with the distance, there will be a mean distance zone where these terms become appreciable. The corresponding energy $E_{m,r.}$ has been evaluated by Margenau (*loc. cit.*) who finds

$$E_{m.r.} = -5.60 \times 10^{-10}\,e^{-5.33r_{AB}}\ \text{ergs}.$$

In conclusion, one may represent the interaction between the two atoms of helium by the expression

$$U = \left[925e^{-4.40r_{AB}} - 560e^{-5.33r_{AB}} - \frac{1.39}{r^6} - \frac{3.0}{r^8}\right] \times 10^{-12}\ \text{ergs}.$$

This purely theoretical expression agrees fairly well with the experimental data. The energy first of all decreases rapidly when the distance increases, passes through a minimum at: $r_{AB} \# 3$ Å (the energy is then about $10^{-15}$ ergs) and then increases again, tending asymptotically towards twice the energy of a helium atom. These results therefore enable us to fix the Van der Waals' radius of helium in the vicinity of 1.5 Å.

One can also see quite clearly that when the interatomic distances are small, the system is in a repulsive state. One may understand this state by observing that it contains two pairs of electrons with the same spins and that the electrons of the same spins will have a tendency to move as far apart as possible, that is to say to locate themselves at the extremities of the system, leaving the nuclei bare, which will then vigorously repel one another. It would be very interesting to calculate the function $\delta$ in this case so as to evaluate the importance of the electronic transfer which must take place from the centre towards the extremities of the system. The nature of the exclusion forces thus appears more clearly. We will observe that, as in the case of hydrogen, the corresponding energy is an exponential function of the internuclear distance with a negative exponent. This remark also applies to the case of two atoms of neon [22] or argon [23].

In the case where the two elements interacting with one another are of a different nature, there may be superimposed on the phenomenon of transfer towards the extremities which involves a repulsion, a transfer of electron going from the less electro-negative system to the more electro-negative system. This transfer of charge renders the domain of the donor positive and the domain of the acceptor negative. The result of this is an attraction which may at least partially compensate for Pauli's exclusion repulsions and produce a stable association at a distance less than the sum of the Van der Waals, radii. The hydrogen bond may however be considered as an extreme case of complex by transfer of charge.

### F. ON FORCES BY CHARGE TRANSFER

It will now be as well for us to give a few brief indications [24] regarding forces by charge transfer. Let D be the donor component and A be the acceptor component.

Mulliken proposed representing the complex by charge transfer with the help of a function of the form

$$\Phi_0 = \alpha \psi_{DA} + b \psi_{D^+A^-}$$

in which $\psi_{DA}$ denotes the function which would represent the system in the absence of transfer and $\psi_{D^+A^-}$ the function which would represent the system if an electron were transferred totally from D to A, D then becoming a positive ion and A becoming a negative ion.

The coefficients $a$ and $b$ may be determined by the application of McDonald's theorem. In this case one also obtains an approximate function

$$\Phi_1 = a' \psi_{DA} + b' \psi_{D^+A^-}$$

which is capable of representing the first excited electronic state of the system.
Since

$$\int \Phi_0^* \Phi_1 \, dv = 0$$

if one ignores the overlap integral between $\psi_{DA}$ and $\psi_{D+A-}$ one will get

$$aa' + bb' = 0$$

whence, because of the relationships bound up with normalisation

$$a' = b \quad \text{and} \quad b' = -a$$

that is to say

$$\Phi_1 = b\psi_{DA} - a\psi_{D+A}.$$

Let us envisage the fairly frequent case in which $a$ is near to unity, so that $b$ is low. This case corresponds manifestly to a low charge transfer in the fundamental state of the system. One can see that on the other hand in the first excited state the function $\Phi_1$ will resemble $\psi_{D+A-}$, that is to say will correspond to a very strong charge transfer. This excited state will therefore have a strong personality and will introduce into the absorption spectrum of the complex a system of bands which belongs neither to D nor to A and which one calls the *charge transfer spectrum*.

The existence of charge transfer forces may therefore be accompanied by specific elements in the electronic spectrum. This is an important experimental criterion for detecting the presence of these interactions.

Hassel and Romming [25] determined with precision the structure of such complexes. In the case of the association $Me_3N, I_2$ which can be represented by the function

$$\Phi_0 = a\Psi_{Me_2N, I_2} + b\Psi_{Me_3N^+, I_2^-}$$

the two atoms of iodine are separated by a distance equal to 2.83 Å and located on the axis of the pyramid $NC_3$. One can therefore see that the transfer has weakened the $I_2$ bond, the length of which is normally 2.67 Å. In this case the distance IN is 2.27 Å, whilst the sum of the Van der Waals' radii is 3.65 Å $(2.5+1.5)$ [26].

## G. THE HYDROGEN BOND

We do not envisage here summarising the quantum theory of the hydrogen bond, which in itself constitutes a vast subject.

We will refer the reader desirous of an analysis of recent studies carried out on this question to the book published by D. Hadzi [27]. We will restrict ourselves to replacing this type of interaction in the assembly of inter-molecular interactions which we are at present analysing.

It is known that the bonds formed by a hydrogen atom and an electronegative element (such as HO, HF ...) interact strongly with other electronegative elements whether or not they are engaged in other molecules.

Thus HF and $F^-$ form the ion $(FHF)^-$ and a OH bond of water attaches itself on the hydrogen side to an atom of oxygen belonging to another molecule.

Furthermore these two molecules correspond to quite distinct types of hydrogen bond. One draws a distinction, in fact, between the long bonds $(R_{AH...B} > 2.7 \text{ Å})$ whose energy is fairly low and the strong bonds $(R_{AH...B} < 2.7 \text{ Å})$ whose energy may reach the order of magnitude of that of ordinary chemical bonds. In the $(FHF)^-$ ion distance FF is 2.25 Å [28] and the bond energy is equivalent to 58 kcal mole$^{-1}$ [29]. It is therefore a strong bond.

The OH...O bond between molecules of water measures 2.76 Å, and it is therefore a weak bond. For such weak bonds the energy varies between 6 and 10 kcal mole$^{-1}$. As the length of the OH bond is of the order of an Å, the length of the OH...O bond would be approximately 3.60 Å$(1 + 1.2 + 1.4)$ if the H and O atoms were separated by the Van der Waals' distance. Generally speaking it is observed that even in the case of the long bonds the interatomic distances fall below those which would correspond to the Van der Waals' association. We are therefore well within the domain of forces with a short radius of action.

We will not analyse the structure of short bonds: the systems which contain them are true molecules. One cannot truly speak any longer of intermolecular forces. We will merely point out that Bessis [30] recently discussed quantitatively the case of the $(FHF)^-$ ion.

On the other hand, the long bonds, even intramolecular as in the case of chelated compounds, must be included in the present chapter. We think it is possible to summarise the essence of the theoretical results by stating that in the hydrogen bond the Pauli exclusion repulsive forces are compensated for by at least three attractive contributions:

(a) the electrostatic forces between the permanent moments of the entities under consideration;

(b) the forces of polarisations;

(c) the charge transfer forces.

The role of the electrostatic forces was emphasised as long ago as 1938 by Bauer and Magat [31]. R. Grahm [32] recently carefully examined the role of the polarisation forces and showed that the corresponding energy has the same order of magnitude as the total energy. Besnainou et al. [33] showed that the introduction of the charge transfer, as envisaged by Sokolov [34], Coulson and Danielson [35] and others, is necessary of one wishes to take quantitatively into account the displacement which affects the bands of the electronic spectrum of a conjugated molecule when it is dissolved in a solvent which is apt to form hydrogen bonds with it. It is assumed that it is the oxygen accepting a hydrogen bond which loses a certain electronic charge in favour of the OH bond. In the method of molecular orbitals this supply of a charge will correspond to the occupation of an antibonding orbital of the O—H bond, which satisfactorily explains the elongation which is experimentally observed for this when it enters into interaction with a hydrogen bond. To sum up, one may write

$$\Phi_0 = \alpha \Psi_{OH...O} + b \Psi_{[OH]^- ...O^+} .$$

We would add that in the case of forces with a short radius of action it is not easy to draw a clear distinction between what is polarisation and what is transfer.

H. STATISTICAL MEANS AND THE RELATIVE IMPORTANCE OF MOLECULAR INTERACTIONS

Let us consider two molecules endowed with permanent electric moments and situated at a relatively great distance.

The interaction energy may be written in the form

$$E = E_{el} + E_{pol} + E_{disp} .$$

We have examined the form of expressions which permit of the calculation of these different terms and we have noted that these depend on the relative orientation of the two molecules under consideration. It is therefore frequently necessary to determine the mean interaction energy $E$, taking into account the different possible orientations.

The probability of meeting two molecules possessing a relative orientation symbolised by $\omega$ will be written

$$e^{-(E(r_{ab}, \omega)/\chi T)}$$

according to Boltzmann's formula if the molecules under study belong to a gas which is in thermodynamic equilibrium.

We are therefore brought to calculate the expression

$$\bar{E} = \frac{\int E(r_{ab}, \omega) e^{-(E(r_{ab}, \omega)/\chi T)} d\omega}{\int e^{-(E(r_{ab}, \omega)/\chi T)} d\omega} .$$

Very often, so long as the pressure is not too high, $E(r_{ab}, \omega)$ more often than not remains weak in relation to $\chi T$. One may then develop in series the exponential function and only retain the first terms of the development. We then get

$$\bar{E}(r_{ab}, T) = \frac{1}{64\pi^6} \int E(r_{ab}, \omega) d\omega -$$

$$- \frac{1}{\chi T} \left[ \frac{1}{64\pi^6} \int E^2(r_{ab}, \omega) d\omega - \left( \frac{1}{64\pi^6} \int E(r_{ab}, \omega) d\omega \right)^2 \right]$$

Margenau [36] and London [37] have shown that, if in this expression one replaces $E$ by $E_{el} + E_{pol} + E_{disp}$, the crossed terms of the type $E_{el}E_{pol}$ or $E_{pol}E_{disp}$, etc., may be ignored in a first approximation, so that one may write

$$\bar{E}(r_{ab}, T) = \bar{E}_{el}(r_{ab}, T) + \bar{E}_{pol}(r_{ab}, T) + \bar{E}_{disp}(r_{ab}, T) .$$

Let us suppose by way of example that the molecules studied are neutral and only endowed with a dipolar moment. By introducing into Expression (1) the formula giving $E_{el}^0$ obtained at the end of Subsection 6B, one finds

$$\bar{E}_{el} = -\frac{2}{3}\frac{\mu_a^2\mu_b^2}{\chi T r_{ab}^6}$$

It is interesting to note that the necessity of taking a mean of the various orientations has converted the original variation which was in $1/r^3$ into a variation in $1/r^6$ and has caused the appearance of a factor $1/T$.

Under the same conditions the formula of Subsection 2C of this chapter leads to

$$\bar{E}_{pol} = -\left[\alpha_b\frac{\mu_a^2}{r_{ab}^6} + \alpha_a\frac{\mu_b^2}{r_{ab}^6}\right].$$

Finally one may also assume a variation in $1/r^6$ for $E_{disp}$ insofar as the molecules are neither too close nor too particular.

In all, the three elements of $E$ commence with a term in $1/r^6$. One may there compare their respective importances independently of this distance. Table VII gives an idea of these importances in the case of interaction between molecules of the same nature at a temperature of 293 K.

According to these examples it is clear that it is difficult to enunciate general rules; one will also observe that the dispersion forces are by far the most important in the case of hydrogen iodide whilst the electrostatic dominate in the case of water.

TABLE VII

| Molecule | $\bar{E}_{el}$ | $\bar{E}_{pol}$ ($10^{-60}$ erg cm$^{-3}$) | $\bar{E}_{disp}$ |
|---|---|---|---|
| CO | 0.003 | 0.06 | 67 |
| HI | 0.35 | 1.7 | 382 |
| $H_2O$ | 190 | 10 | 47 |

I. EMPIRICAL TREATMENT OF INTERMOLECULAR INTERACTIONS

We are now in a position to understand the origin of the semi-empirical potentials which have been proposed for representing intermolecular interactions and we are in a position to use them deliberately. In the case of two non-polar molecules we have seen that the forces with a long radius of action all lead to the introduction of a term in $-1/r^6$.

We have noted that the forces with a short radius of action, on the other hand, involve the appearance of an exponential term.

Buckingham's potential

$$E = be^{-ar_{ab}} - \frac{c}{r_{ab}^6} - \frac{c'}{r_{ab}^8}$$

therefore possess a form which is well adapted for representing interactions between such molecules*.

Lennard-Jones proposed a function of the form

$$E = \frac{d}{r^\delta} - \frac{c}{r^\gamma}.$$

If one takes $\gamma = 6$ and $\delta = 12$ one obtains an expression fairly close to that of Buckingham because the term in $1/r^{12}$ has a behaviour which is close to the exponential except at very small distances where it just avoids the drawback of Buckingham's potential.

In cases where it is desirable to use semi-empirical potentials for a given orientation of the molecules, one may make use of Stockmayer's function

$$E = 4\varepsilon \left[ \left( \frac{\sigma}{r} \right)^{12} - \left( \frac{\sigma}{r} \right)^6 \right] - \frac{\mu_a \mu_b}{r^3} (2 \cos \theta_a \cos \theta_b - \sin \theta_a \sin \theta_b \cos \varphi).$$

It is clear that it will be advisable to choose other types of potential as soon as one moves away from the conditions which we have just stipulated and which are the only ones for which the formulae we have just indicated are suitable.

### 3. Interaction Between Unbonded Atoms

A. VAN DER WAALS' RADII

From the examination of the structure of a paraffinic crystal one can see the fact that the shortest distance between hydrogen atoms belonging to the two distinct molecules is of the order of 2.48 Å. It is usual to call one-half of this distance (or 1.24 Å) the Van der Waals' radius of hydrogen. Generally speaking the Van der Waals' radius of an atom corresponds to the minimum of energy of interaction of this atom with an unbonded atom of the same nature.

Table VIII [38] gives the values usually accepted for the Van der Waals' radii.

TABLE VIII

| H 1.24 | | | | He 1.5 |
|--------|------|---------|---------|--------|
| C 1.7  | N 1.5 | O 1.4  | F 1.35  | Ne 1.6 |
|        | P 1.9 | S 1.85 | Cl 1.80 | A 1.9  |
|        | As 2.0 | Se 2.00 | Br 1.95 | |

Consequently, in principle one must expect that the interactions between two atoms will be repulsive if their distance is less than the sum of their Van der Waals' radii and attractive otherwise.

It is clear, however, that this rule results from a simplification and that a great deal

---

* Nevertheless one must note that this potential is incorrect at very short distances, because it then tends towards $-\infty$.

of caution is necessary in applying it, particularly in the case where the atoms under consideration are concerned in bonds and for this reason highly perturbed.

The notion of a Van der Waals' radius is particularly of interest for qualitative discussions in cases where one is *considerably* below or above the critical distance.

Let us consider, for example, the case of benzene, where the distance which separates two carbon atoms in a meta position in relation to one another is 2.4 Å.

This distance is very much less than the sum of the Van der Waals' radii (3.4 Å). Repulsive forces must therefore intervene between these two atoms. Now let us consider the case of phenanthrene

The two hydrogen atoms set out in the preceding formula would be separated by a distance of approximately 1.7 Å if they remained in the plane of the molecule. It is therefore probable that repulsive forces would compel these atoms to move out of this plane. Trotter [39] appears to have shown that this is in fact the case in the analogous instance of benzbisanthrene

This preamble shows very clearly the very general importance of interactions between unattached atoms.

### B. CONFORMATIONS OF ACYCLIC PARAFFINS, POTENTIAL BARRIERS TO INTERNAL ROTATION

The interactions between unattached atoms intervene in the simplest molecules. To convince oneself of this it is sufficient to recall, for example, that the distance between two hydrogen atoms in methane is 1.8 Å. Let us now take the case of ethane and let us envisage the so-called 'free' rotation of one of the methyl groups in relation to the other. During this rotation the interactions between unattached atoms will manifestly vary because certain distances H, H vary. The periodic properties of the phenomenon lead to the interaction energy $E$ developing in a Fourier series. One may therefore write

$$E = \sum_{n=1}^{\infty} \tfrac{1}{2} V_n (1 - \cos n\varphi)$$

if $\varphi$ signifies the angle between two planes containing the CC axis and one of the CH bonds of each methyl.

The determination of the values of $V_n$ by means of wave mechanics is still very difficult. Parr presented an important communication on this subject to the summer school which we organised at Menton in July 1965. In practice one therefore restricts oneself to finding the values of $V_n$ which are most the appropriate for accounting for the spectroscopic data and in particular those resulting from the study of the micro-wave spectra.

In the case of methylchloroform it has been possible in this way to take two terms into account and find [40]

$$V_3 = 2910 \quad \text{and} \quad V_5 = 57 \text{ cal mole}^{-1}.$$

Very frequently one is content to use one term only and consequently to assume that

$$E = \tfrac{1}{2}V_n(1 - \cos n\varphi)$$

$n$ in each case being suitably chosen.

One then calls the value obtained for $V_n$ the *potential barrier to internal rotation*. Table IX [41] contains a few values of this magnitude.

TABLE IX

| Molecule | $n$ | $V_n$ (cal mole$^{-1}$) |
|---|---|---|
| $H_3C - CH_3$ | 3 | $2875 \pm 125$ |
| $H_3C - CF_3$ | 3 | $3000 \pm 200$ |
| $H_3C - C_2H_3$ | 3 | 1950 |
| $H_3C - CHO$ | 3 | 1000 |
| $H_3C - NO_2$ | 6 | $6 \pm 0.3$ |

Figure 10a shows the variation of $E$ with the angle $\varphi$ in the case of ethane as well as the conformations corresponding to the minima and maxima of $E$. Figure 10b corresponds to the case of normal butane.

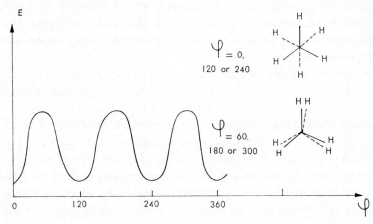

Fig. 10a.

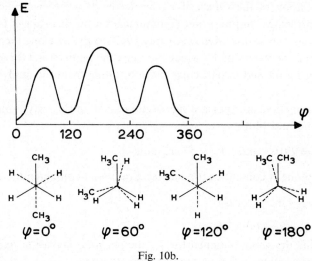

Fig. 10b.

When a classically single bond is adjacent to a classically multiple bond or situated between two such bonds it undergoes a certain shortening.

Table X [42] emphasises this fact on the basis of recent determinations.

TABLE X

| Molecule | Bond | Length | Shortening |
|----------|------|--------|------------|
| Propylene | $C-C$ | 1.50 | 0.02 |
| Octyl cyanide | $C-C$ | 1.49 | 0.03 |
| Vinyl chloride | $C-Cl$ | 1.72 | 0.05 |

This phenomenon is usually explained by introducing the idea of hyper-conjugation or conjugation or again by taking into account the variation in the hybridization which characterises the orbitals responsible for the bonds when changing from a single bond surrounded only by single bonds to a single bond which is adjacent to one or more multiple bonds.

However, certain authors [43] and [44] have put forward the view that it is the forces between unattached atoms which are in fact the principal cause of the phenomenon.

They assume that the apparent shortening of the single bonds which have just been discussed is due to the fact that a single bond adjacent to other single bonds (as in the case of ethane) is in a state of elongation because of the strong repulsive forces which occur between unattached atoms and that when one replaces one of the single bonds adjacent to the bond under study by a multiple bond, one reduces the number of un-bonded atoms and consequently the repulsion. The shortening observed would therefore in fact be a decrease in an elongation.

E. B. Wilson (*loc. cit.*), however, has given arguments against this interpretation. He observes in particular that the potential barriers to internal rotation of the derivatives of ethane do not vary much when one replaces a hydrogen atom by another atom, as can be seen from Table XI. From this he deduces that it is not very probable that under these conditions a strong repulsion can exist between the hydrogen atoms of a molecule like ethane which is likely to produce an appreciable elongation of the CC bond.

TABLE XI

| Molecule | Barriers (kcal mole$^{-1}$) |
|---|---|
| $CH_3CH_3$ | 2.8 |
| $CH_3CH_2F$ | 3.3 |
| $CH_3CHF_2$ | 3.2 |
| $CH_3CH_2Cl$ | 3.56 |
| $CH_3CH_2Br$ | 3.57 |
| $CH_3CH_2CN$ | 3.5 |

Whatever may be the case, the knowledge of the function $E(\varphi)$ permits of a knowledge of the energies of the different *conformations* of the molecule, that is to say of the structures corresponding to the different energy minima.

It may be considered that each of these conformations constitutes a chemical species. Since the barriers to internal rotation are weak at ordinary temperatures, these different chemical species are usually in equilibrium. Pitzer [45] has produced tables which make it possible to calculate the characteristic thermodynamic magnitudes of such molecules on the basis of the data analysed in this section.

In principle therefore one could deduce from the energy of the different conformations the equilibrium constants which connect them and in this way obtain a cross-check with the structural spectroscopic data because these constants can be measured by physical methods of analysis (infrared, Raman). This cross-check is still rather risky at the present time because of the inaccuracies which usually exist in our knowledge of $E(\varphi)$ and because of the necessity of taking into account all the vibration/rotation movements. Nevertheless, in Subsection 3C of this chapter we shall be analysing in detail a problem which is fairly close to this.

C. CONFORMATIONAL EQUILIBRIA IN THE CYCLANE SERIES, BAEYER'S TENSION,
   COTTON'S EFFECT

In principle the methods for studying the inter-molecular forces which we have examined in Section 2 of this chapter should help us to calculate theoretically the energy of the molecular conformations and open up the way to predicting conformational equilibrium constants.

Unfortunately we have seen that the energy of interaction between unbonded atoms varies abruptly with the distance slightly above the sum of the Van der Waals' radii and small errors in estimating this critical distance can give rise to great in-accuracies in regard to the value of the energy calculated for a conformation. Calcula-tions *ab initio* are therefore not very advisable. It is preferable to make use of semi-em-pirical processes.

So as to demonstrate with a concrete example how it is possible to operate, we will study a contribution to the conformational analysis in the cyclopentane series based on calculations of these types and on the application of the *octant rule* which makes it possible to connect with the geometry of a conformation the sign of its *Cotton effect*. It may perhaps be as well to recall briefly what the Cotton effect and the octant rule consist of.

It is known that Biot [46] observed in 1817 that the power of rotation of a substance depends on the wavelength of the polarised light which passes through it. This fact is interpreted by noting that any rectilinearly polarised wave may be considered as the result of the superimposition of a left circular wave and a right circular wave and that these two waves are not propagated at the same velocity. If one uses $n_L$ to indicate the index of the medium for the left wave and $n_R$ to designate the index of the medium for the right wave, one immediately obtains the formula

$$\alpha = \frac{\pi}{\lambda} (n_L - n_R)$$

if $\alpha$ designates the rotatory power in radians per unit length. The specific rotation is defined by

$$[\alpha] = \frac{\alpha}{c'} \frac{1800}{\pi}$$

if the unit employed is one degree per decimetre and $c'$ denotes the concentration of the molecules studied. The formula

$$[\Phi] = [\alpha] \frac{M}{100}$$

finally connects the molecular power of rotation with the specific rotation.

It should be observed that a very slight difference between the two indices $n_L$ and $n_R$ is sufficient to bring about the existence of a very appreciable power of rotation. Thus for

$$n_L - n_R = 3 \times 10^{-6}$$
$$\text{and} \quad \lambda = 5500 \text{ Å}$$

one gets

$$\alpha = 10° \text{ cm}^{-1}.$$

The presence of the rotatory power is accompanied by circular dichroism in that the difference between the indices corresponds to a difference in the molecular extinc-tion coefficients $\varepsilon_L$ and $\varepsilon_R$ of the left and right waves. As the left component of the wave is thus absorbed with a speed which differs from that which characterises the right component, a certain ellipticity appears in the resultant wave.

If one measures this ellipticity by the angle $\theta$, the tangent of which is equal to the ratio between the short axis and the long axis of the ellipse, one obtains the equation

$$\theta = \tfrac{1}{4}(k_L - k_R)$$

if the values of $k$ signify the absorption coefficients of the two waves and if one introduces the molecular ellipticity

$$[\theta] = \theta \frac{18}{\pi} \frac{M}{c'}$$

one finally reaches the formula

$$[\theta] = 2.30 \left(\frac{4500}{\pi}\right)(\varepsilon_L - \varepsilon_R).$$

The existence of different speeds and absorption coefficients for right and left waves is given the name of Cotton's effect [47].

As the ellipticity is generally very slight, one prefers to study the power of rotation.

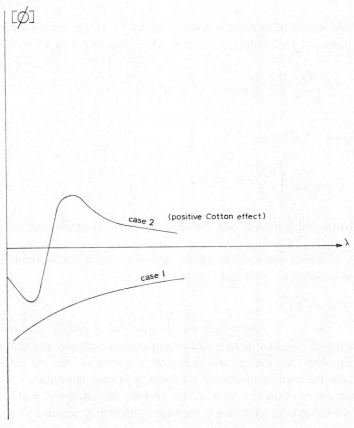

Fig. 11.

The curve representing the variation of the rotary power of a given substance with the wavelength $\lambda$ may be either monotonous or, on the other hand, it may comprise a sort of wave motion which passes from values of one sign to values of the opposite sign (Figure 11).

When the curve shows a wave motion passing from negative values to positive values it is said that the molecule presents a positive Cotton effect. It is said to correspond to a negative Cotton effect if the wave motion has its positive part before its negative part.

A detailed analysis of the phenomenon has shown that such a wave motion always appears in the vicinity of an absorption band.

It is also known that according to Kronig [48] and Kramers [49] a knowledge of the law of absorption in terms of wave length makes it possible to determine the curve of the corresponding dispersion.

When applied to the case of the Cotton effect, this remark leads to the formula [50]

$$[\Phi(\lambda)] = \frac{2}{\pi} \int\limits_0^\infty [\theta(\lambda')] \frac{\lambda'}{\lambda^2 - \lambda'^2} \, d\lambda'$$

connecting the power of rotation and the ellipticity for an absorption band. Likewise, when one defines the oscillatory force of an absorption band by the equation

$$f = \frac{3hc}{8\pi^3 N} \int\limits_0^\infty \frac{k(\lambda)}{\lambda} \, d\lambda \approx 0.92 \times 10^{-38} \int\limits_0^\infty \frac{\varepsilon(\lambda)}{\lambda} \, d\lambda$$

one may define the rotational force

$$R = \frac{3hc}{8\pi^2 N} \int\limits_0^\infty \frac{\theta(\lambda)}{\lambda} \, d\lambda \approx 0.696 \times 10^{-42} \int\limits_0^\infty \frac{[\theta(\lambda)]}{\lambda} \, d\lambda$$

and the relation which links $\Phi$ with $\theta$ makes it possible in simple cases to link $\Phi$ with $R$ [51].

Just as $f$ directly linked with the transition moment associated with the band under study, it has been shown [52] and [53] that

$$R = \vec{\mu}_e \cdot \vec{\mu}_m$$

in which $\vec{\mu}_e$ and $\vec{\mu}_m$ represent induced electric and magnetic dipolar moments.

It is then found [54] that if the molecule possesses a centre or plane of symmetry, $R$ is generally zero (because one of the moments is zero or they are perpendicular), which explains the absence of a power of rotation in such molecules.

Let us now consider the case of a cyclane ketone. The band $n \to \pi$ of the carbonyl group, the wavelengths of which are in the vicinity of 2900 Å, appears to be electrically suppressed ($\vec{\mu}_e = 0$) but magnetically permitted ($\vec{\mu}_m \neq 0$) if one ignores the perturbations due to the environment of the carbonyl. It results from this that in this approximation

$$R = 0.$$

In point of fact, an asymmetry in the environment of the group CO may cause the appearance of a certain moment $\vec{\mu}_e$ and therefore finally that of a certain rotational force. The measurement of the Cotton effect therefore gives for the type of substance envisaged a direct indication regarding the nature of the environment of the CO group.

To describe the environment one usually considers the eight octants formed by the plane $P_1$ containing the CO bond and the two adjacent C—C bonds, the plane $P_2$ containing the CO bond and perpendicular to $P_1$ and the plane $P_3$ passing through the oxygen and perpendicular to $P_1$ and $P_2$. Figure 12 (a) shows the nature of these planes and Figure 12 (b) shows the projection on the plane of $P_3$ the different atoms of the molecule and states the nomenclature used for designating the octants. Now it has been observed empirically that the substituents carried by the atom 4 and situated in $P_2$ have no action. The same applies to those connected to $D_2$ and $G_2$ and situated in the plane $P_1$.

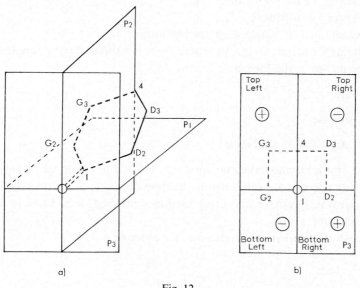

Fig. 12.

The substituents connected to $D_2$ situated in the 'bottom right' octant and the substituents connected to $G_3$ and situated in the 'top left' octant give a positive contribution to the effect. The substituents attached to $D_3$ situated in the 'top right' octant and the substituents attached to $G_2$ in the 'bottom left' octant give a negative contribution to the Cotton effect. *This is the octant rule.*

But the practical application of this rule is not always easy because at ordinary temperature the cyclanones may take the form of a mixture of different conformations and the Cotton effect observable results from a superimposition of the contributions of each conformation. It is therefore necessary to know the relative concentrations of the different conformations. One means of obtaining an estimate of these magnitudes is to calculate their energy and to apply Boltzmann's law. C. Ouannes [55] has studied the case of cyclopentanones. We will summarise this study. For cyclopentane one may consider two principal conformations: the envelope conformation

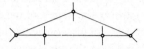

and the half-chair conformation

The calculation of the energy of the conformations may be carried out by taking into account successively:

(a) the energy of Baeyer's tension $E_B$;
(b) the energy of torsion $E_T$;
(c) other interactions between unattached atoms.

By restricting oneself to the harmonic force constant, Baeyer tensions may be assessed by means of the formula

$$E_T = K_0 (\theta - \theta_0)^2.$$

C. Ouannes chose

$$K_0 = 0.0175 \text{ kcal mole}^{-1}$$

$\theta_0 = 109.47°$ for a tetrahedral carbon and $\theta_0 = 116.3°$ for a trigonal carbon.

The torsion energies are calculated by treating each side of the pentagon as a substituted derivative of ethane and using the functions $E(\varphi)$ introduced in Subsection 3B of this chapter.

In the case of 3-methylcyclopentanone, for example

the torsion energy around the bond 4 is calculated from the function $E(\varphi)$ of ethane, the energy relating to bond 2 from the function of $CH_3CH_2CH_3$ and that corresponding to the bond 1 from the function of $CH_3CHO$.

It remains to introduce the other interactions between unattached atoms, that is to say those which are exerted between atoms attached to non-adjacent carbons.

C. Ouannes adopted the semi-empirical formula of Hill [56]

$$E_{n \cdot l} = 8.28 \times 10^5 \; \varepsilon \; e^{-(r/0.0736r*)} - \frac{2.25 \; \varepsilon r^{*6}}{r^6}$$

in which $\varepsilon$ is a coefficient which depends on the atoms studied and in which $r^*$ denotes the sum of their Van der Waals' radii.

In this one can recognise an expression of the Buckingham type (Subsection 2I of this chapter).

Figure 13 contains the values of the energies calculated in this way for the conformations of 3-methylcyclopentanone selected as an example. It can clearly be seen that the effect of conformation no. 2 must compensate for that of conformation no. 1.

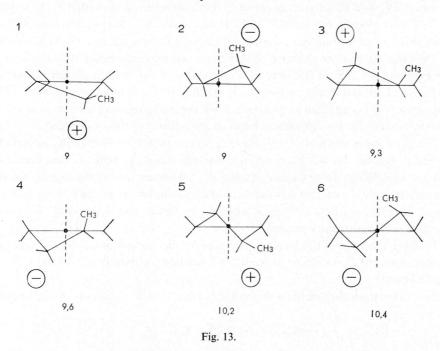

Fig. 13.

On the other hand, the effect of conformation no. 3 prevails over that of conformation no. 4. Likewise, the effect of conformation no. 5 prevails over that of conformation no. 6. Since conformations nos. 3 and 5 correspond to a positive Cotton effect, one can conclude from this that the Cotton effect of this cyclopentanone must be positive. This is in fact what we learn from experimental data. Naturally this type of reasoning is only indicative. A better discussion would require a knowledge of the values of the Cotton effect associated with each conformation.

D. TOTAL ENERGY OF CONJUGATED MOLECULES, STERIC EFFECTS, SPATIAL CONJUGATION

A study of the conformations of conjugated molecules is seen to be rather more complex than that relating to saturated molecules because the rotations around the single bonds are very frequently accompanied by a perturbation of the energy of the delocalised bonds which reinforce them.

It therefore seems to us to be a good plan to take up first of all an examination of the calculation of the total energy of conjugated molecules, a question which has made quite appreciable progress since the publication of the second book of this series.

Chung and Dewar [57] have developed a process which makes it possible to calculate with a good degree of accuracy the energy of formation $\Delta H^T$ of an aromatic hydrocarbon.

This process starts off from the formula

$$- \Delta H_f^T = N_C D_{CC}^T + N_A D_{CH}^T + E_d$$

in which $D_{CC}^T$ and $D_{CH}^T$ represent respectively the energies of formation at the temperature $T$ of the localised bonds CC and CH and in which $E_d$ denotes the energy of the delocalised bond (taking into account its interaction with the rest of the molecule).

The calculation of the energy $E_d$ has been carried out using either Hückel's method (second book, p. 165) or the method of the auto-coherent field in the approximation of Pariser, Parr and Pople (second book, p. 205), or the method of *split orbitals* [58-60] which may be regarded as a variant of the preceding method, consisting in taking into better account the correlation between the positions of the electrons*.

The parameters which are useful for these calculations have formed the subject of a careful discussion. We will point out, for example, that three methods have been tried out for calculating the resonance integral $\beta_{CC}$. When examining the case of benzene the authors have shown that it is the method consisting in estimating $\beta$ in terms of the length of the bonds by a technique proposed by Dewar and Schmeising [61] which gave the most satisfactory results.

To check the value of the formula proposed for the calculation of $\Delta H_f^T$, Chung and Dewar evaluated $E_d$ for eleven aromatic hydrocarbons, assuming equal lengths for all the CC bonds.

This calculation then permits of the estimation of $D_{CC}^{298}$ by means of the formula

$$D_{CC}^{298} = \frac{1}{N_C} (- \Delta H_f^{298} - N_H D_{CH}^{298} - E_d)$$

because $\Delta H_f^{298}$ can be deduced from the experimental values of the combustion heats collected by Wheland [62] and $D_{CH}^{298}$ may be estimated empirically at 102.13 kcal mole$^{-1}$.

Table XII contains a few values thus obtained. Column I corresponds to the case where $E_d$ is calculated by Hückel's method, column II corresponds to the case where $E_d$ results from the use of the method of Pariser, Parr and Pople and column III is derived from the method of split orbitals.

It will be seen that the values calculated for $D_{CC}^{298}$ do not depend a great deal on the method and that moreover for one and the same methods these values vary very little when one passes from one molecule to another molecule.

* Which is in any case the opinion of the authors.

TABLE XII

| | $D_{CC}^{298}$ (in eV) | | |
| | I | II | III |
| --- | --- | --- | --- |
| Benzene | 4.079 | 3.832 | 3.998 |
| Naphthalene | 4.066 | 3.831 | 3.997 |
| Pyrene | 4.071 | 3.837 | 4.006 |
| Chrysene | 4.064 | 3.824 | 3.993 |
| Perylene | 4.053 | 3.817 | 3.987 |

It does seem that one can assume that the energy of a CC or CH single bond is equal to a constant for all the aromatic hydrocarbons.

Armed with this result, Chung and Dewar attacked a more difficult problem by trying to foresee the nature of the structures endowed with an *aromatic character*. More precisely, they tried to see whether their method of calculation of the total energy of conjugated molecules makes it possible to predict the geometry of a cyclic polyene $(CH)_n$ which is still referred to as an annulene.

It is known that *Hückel's rule* [63] suggests that the condition for an annulene to be aromatic is

$$n = 4m + 2$$

$m$ being an integer. Numerous experimental facts are in agreement with this rule. The best known among them reside in the fact that benzene is plane whilst cyclooctatetraene is not. Jackman *et al.* [64] have also recently studied the magnetic behaviour of numerous cyclic polyenes and have observed that this was in agreement with Hückel's rule.

But Hückel did not develop any convincing theoretical arguments to serve as a basis for his rule. Dewar [65] carried out very approximate calculations in order to support it. Nevertheless, it still remained for the problem to be treated in a more precise fashion: this is what Chung and Dewar have done.

Their method consists in evaluating the resonance energy $E_R$ of the annulenes defined as the difference between the energy of the non-aromatic structure where the double bonds alternate with single bonds and the energy of the structure containing a delocalised bond extending over the entire skeleton.

If this energy is positive the more stable aromatic structure is called for, otherwise it is the alternating structure which has to be preferred.

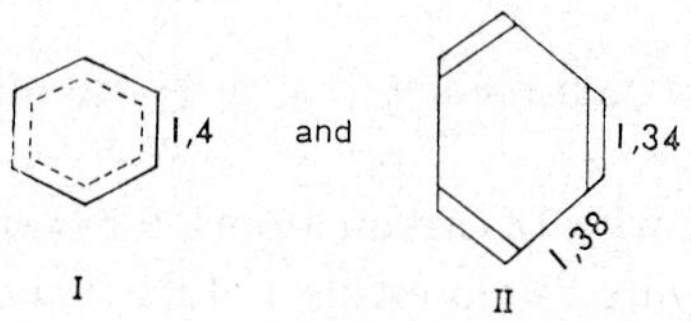

Fig. 14a.

In the case of benzene one will therefore have to calculate the energy of the two structures of Figure 14a.

One may therefore write

$$R_R = E_l^I - E_l^{II}$$

if the $E_l$ represent the bond energies.

The results obtained when studying $\Delta H_f^T$ enable us to write

$$E_l^I = N_C E_{CC} + N_H E_{CH} + E_d$$

in which $E_{CC}$ and $E_{CH}$ represent the energies of the CC and CH bonds which, as we have seen, may be considered as more or less independent from the hydrocarbon. In any case we know how to calculate $E_d$.

On the other hand one will have

$$E_l^H = n'E' + n''E'' + N_H E_{CH}$$

if $n'$ is the number of single bonds of a length 1.48 Å of energy $E'$ and $n''$ is that of the double bonds of energy $E''$.

From this we get

$$E_R = (N_C E_{CC} - n'E' - n''E'') + E_d$$

which one can also write

$$E_R = n(A_0 + a_1 A_1) + E_d$$

by inserting

$$a_1 = \frac{1}{n}\left(n' - \frac{n}{2}\right)$$
$$A_0 = E_{CC} - \tfrac{1}{2}(E' + E'')$$
$$A_1 = E_{CC} - E'.$$

$A_0$ and $A_1$ have been treated as empirical parameters using hydrocarbons for which the resonance energy is known. Figure 14b shows the results obtained by the method used for calculating $E_d$. It can clearly be seen that only the method of split orbitals gives interesting results demonstrating Hückel's rule up to $n = 12$.

Nevertheless, this result is not entirely satisfactory, because according to Jackman et al. (loc. cit.) annulene with 22 carbon atoms would still be aromatic.

This is the reason why Dewar and Gleicher [66] carried out their calculations again by introducing the differences which exist between the lengths of the CC bonds in conjugated hydrocarbons.

But when one introduces such refinements, there are other peculiarities which can no longer be ignored.

Let us consider annulene with 18 carbon atoms, for example; numerous conformations may be envisaged. Figure 15 represents 8 of these and Dewar and Gleicher have shown that the energy of the delocalised bond depends a great deal on the conforma-

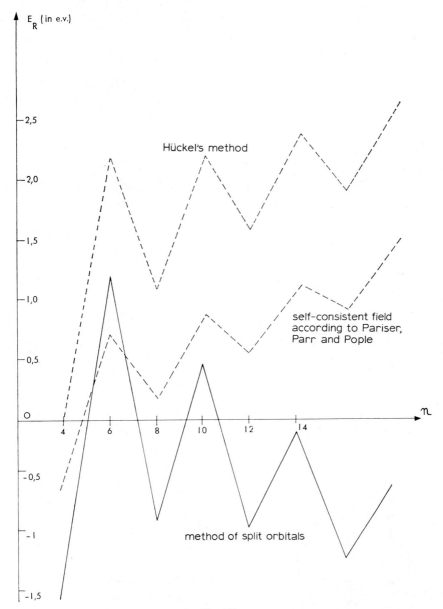

Fig. 14b.

tion chosen. However, it is very difficult to calculate correctly the total energies of these conformations. In this way in the case of conformation I the repulsions between the hydrogen atoms situated inside the cycle would introduce angular deformations and terms of energy which would be very tricky to estimate.

For this reason Dewar and Gleicher have preferred to base their arguments on experimental data (X-ray diffraction) which show that the most open conformation

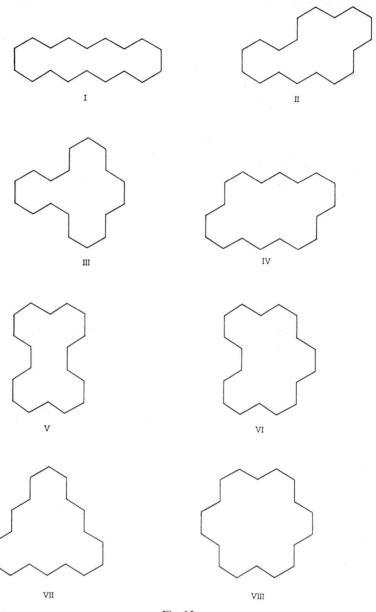

I

II

III

IV

V

VI

VII

VIII

Fig. 15.

is preferred by the annulene under consideration. They have extrapolated this conclusion and calculated the resonance energies by an *improved* method of fractionated
orbitals for the most open conformations of each annulene.

Figure 16 shows the result of the calculations. This time theory and experiment seem
to agree remarkably. In point of fact, always according to Jackman *et al.*, Hückel's

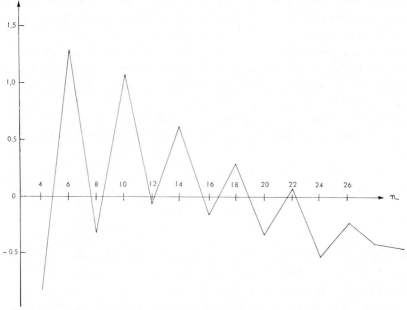

Fig. 16.

rule would apply up to $n=22$ but would not be valid any longer for $n=30$; the corresponding annulene not being aromatic.

Now it is in fact beyond $n=22$ that, according to the curve of Dewar and Gleicher, the aromatic character definitely seems to disappear. It should be observed that, as we have seen (second book, p. 184), Longuet-Higgins and Salem [67] and other authors had predicted that from a certain value of $n$ onwards the aromaticity would be replaced by an alternation of weak and strong bonds, but it had not been possible to state precisely the particular value of $n$.

We will also point out that Dewar and Gleicher have calculated the resonance energies of numerous non-alternant hydrocarbons. They find that these are weak as compared with those which correspond to alternant hydrocarbons. Azulene shows the strongest resonance energy in the series under study. They also note that according to the bond indices a marked alternation of strong and weak bonds should be observed in these molecules (with the exception of azulene) so that it would be better not to class these substances in the aromatic series. Finally, in a third article in the same series, Dewar and Gleicher have examined the case of acyclic polyenes. They have observed that for these molecules the formation energy may be obtained by addition by assuming the alternation of almost single bonds and almost double bonds. They conclude that in no case can one attribute the aromatic character to these substances.

In conclusion it may be said that if these works are correct, the *only ones which would be aromatic among the molecules examined here would be the annulenes of the type $4m+2$ for $m \leqslant 5$ and molecules resulting from the joining of such annulenes, in* *particular the condensed aromatic hydrocarbons and also azulene.* It is obvious that we will have to take the greatest possible account of these considerations during the study of the chemical reactivity of these substances.

It remains for us now to take into account more explicitly certain interactions between unattached atoms. We will examine certain cases of steric hindrance and certain phenomena of space conjugation. We have already pointed out cases of steric encumberment in the introduction to this section. In this way we have called to mind the case of certain of the hydrogen atoms of phenanthrene. *A priori* the repercussions of an encumberment of this type can be very numerous. Let us take the case of diphenyl

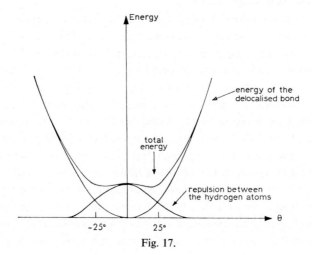

The hydrogen atoms fixed at 2 and 2', for example, would be at a distance (1.78 Å) which is distinctly less than the sum of the Van der Waals' radii if one assumed a classic and plane structure for this molecule.

The consequences of this encumberment must give rise to at least five types of distortions:

(1) the elongation of the 1-1' bond;

(2) the deformation of the $C_2H$ and $C_2'H$ bonds in the plane of the molecule;

(3) the deformation of these same bonds outside this plane;

(4) the shortening of these same bonds;

(5) the rotation $\theta$ of the phenyl groups in relation to one another around the 1-1' bond.

Coulson [68] has shown that of all these effects it is the last-mentioned which is the most important. Figure 17 shows the result of the calculation of the energy of di-

Fig. 17.

phenyl as a function of the angle $\theta$ according to this memorandum and also that of Adrian [69]. The minimum value of the energy would be in the vicinity of 25°. This fact agrees qualitatively with the measurements of Almenningen and Bastiansen [70] who, with the help of the diffraction of the electrons on the gas observed a libration around $\theta = 41$. The fact that this angle exceeds the calculated value seems to indicate that the repulsion between the hydrogen atoms has been underestimated in Adrian's calculation.

The curve of Figure 17 also shows that the energy does not vary very much with $\varphi$ between 0 and 25°. One can deduce from this that diphenyl must easily tend towards planeity under the influence of pressure. This is doubtless why in the crystallised state its molecules are plane. One will take good note of this possibility of obtaining for one and the same molecule different conformations according to whether it is in the gaseous phase or in the condensed phase.

Let us add that Adrian (*loc. cit.*) also discussed the case of triphenylmethyl.

We will recall on the other hand that we have already referred to the works of Coulson and Senent who examined the structure of 3,4-5,6-dibenzophenanthrene (second book, p. 187) and showed that in this case the steric encumberment between the two atoms of hydrogen brings about a sort of warping of the whole of the molecule, which thus presents a slightly helicoidal structure.

Coulson and Golebiewsky [71] studied the consequences of the steric hindrance which arises between the internal hydrogen atoms of the annulenes and concluded that these atoms must move away from the plane of the molecule, placing themselves alternately on one side or other of the plane.

Coulson and Stocket [72], Haigh [73], and Hendrickson [74] have developed a general method for dealing with this type of problem.

To end this chapter which is mainly devoted to the interactions between unattached atoms, it remains for us to examine the nature and the properties of the phenomena of *spatial conjugation*.

Let us consider for example norbornylene (I) and bicycloheptadiene (II).

I                                                    II

One would expect that the electronic spectra of these two molecules would be located in the same region of wavelength, the compound (II) containing simply the double bond of the compound (I) twice over.

Experiments show us that this is not the case; the band of the greatest wavelength of norbornylene corresponds to a transition energy of 6.33 eV. The spectrum of bicycloheptadiene possesses a band at 5.85 eV and another band at 6.87 eV [75]. The nett bathochromic effect has been interpreted by assuming that the two double bonds of bicycloheptadiene interact and that it is advisable to introduce a delocalised bond with four centres in this molecule, as suggested by the following scheme

It is sometimes said that there is *spatial conjugation* because the delocalised bonds extend between atoms which are classically unattached.

Wilcox *et al* [76] and then Hermann (*loc. cit.*) made use of the approximation of Pariser and Parr to analyse this phenomenon quantitatively. We will summarise the results of Hermann's calculations which are a little more complete than those of his predecessors.

The table below gives the values of the parameters adopted:

$$
\begin{aligned}
(11/11) &= 11.08 \ \text{eV} & \beta_{12} &= -2.70 \ \text{eV} \\
(11/22) &= 9.36 & \beta_{13} &= 0 \\
(11/33) &= 5.239 & \beta_{14} &= 0 \\
(11/44) &= 6.059 &
\end{aligned}
$$

The parameter $\beta_{12}$ was chosen so as to obtain an agreement between theory and experiment in the case of norbornylene. The other integrals $\beta$ were taken to be equal to zero. The formulae of Pariser and Parr [77] lead effectively to very low values for these parameters ($\beta_{14} = 0.0178$ and $\beta_{13} = 0.0031$).

The coulombic integrals $(ii/jj)$ were calculated by the usual processes, taking into account the non-zero angle which is formed between the axes of the orbitals $\pi$. The calculation of the energies of the electronic levels was carried out by means of the method of configuration interactions, using practically all the mono and bi-excited configurations. See Table XIII for the principal results.

TABLE XIII

| Nature of transistion | Calculated energy | Energy determined by experiment |
|---|---|---|
| $1_{A_1} \rightarrow 1_{A_2}$ | 5.495 | 5.85 |
| $1_{A_2} \rightarrow 1_{B_2}$ | 7.134 | 6.87 |

One may therefore conclude that the calculations based on the interpretation proposed here give very satisfactory results.

Numerous other examples of spatial conjugation have formed the subject of theoretical studies.

We may mention the case of the paracyclophanes [78], germacrol and barralene [79].

The three-dimensional conjugation phenomena observed in the series of the phenylated acene derivatives [80] present themselves in a more specialised light and pose specific problems, the examination of which has just been commenced [81].

## References

[1] R. J. Buehler and J. O. Hirschfelder, *Phys. Rev.* **83** (1951) 628; **85** (1952) 149.
[2] See for example L. Pauling and E. B. Wilson, *Introduction to Quantum Mechanics*, McGraw-Hill, New York, 1935, p. 204.
[3] J. C. Slater and J. G. Kirkwood, *Phys. Rev.* **37** (1931) 682.
[4] H. R. Hassé, *Proc. Camb. Phil. Soc.* **27** (1931) 66.
[5] L. Pauling and J. Y. Beach, *Phys. Rev.* **47** (1935) 686.
[6] H. Margenau, *Phys. Rev.* **38** (1931) 747.
[7] F. London, *Z. Phys.* **63** (1930) 245.
[8] J. F. Hornig and J. O. Hirschfelder, *J. Chem. Phys.* **20** (1952) 1812.
[9] H. B. G. Casimir and D. Polder, *Phys. Rev.* **73** (1938) 360.
[10] F. London, *J. Chem. Phys.* **46** (1942) 305.
[11] P. L. Davies and C. A. Coulson, *Trans. Far. Soc.* **48** (1952) 777.
[12] H. C. Longuet-Higgins and L. Salem, *Proc. Roy. Soc.* A**259** (1961) 433.
[13] A. G. de Rocco and W. G. Hoover, *Proc. Nat. Acad. Sci.* **46** (1960) 1057.
[14] L. Salem, *Nature* **193** (1962) 476
[15] L. Salem, *Can. J. of Biochem. Phys.* **40** (1962) 1287.
[16] W. Kolos and C. C. J. Roothaan, *Rev. Mod. Phys.* **32** (1960) 219.
[17] H. M. James, A. S. Coolidge, and R. D. Present, *J. Chem. Phys.* **4** (1936) 187, 193.
[18] J. O. Hirschfelder and J. W. Linnett, *J. Chem. Phys.* **18** (1950) 130.
[19] P. Rosen, *J. Chem. Phys.* **18** (1950) 1182.
[20] H. Margenau, *Rev. Mod. Phys.* **11** (1939) 1.
[21] C. H. Page, *Phys. Rev.* **53** (1938) 426.
[22] W. E. Bleick and J. E. Meyer, *J. Chem. Phys.* **2** (1934) 252.
[23] M. K. Kunimune, *Prog. Theor. Phys. Japan* **5** (1950) 412.
[24] The reader may find important literature on this subject in the following works or articles: G. Briegleb, *Elektronen-Donator-Acceptor-Komplexe*, Springer Verlag 1961; R. S. Mulliken and W. B. Person, *Ann. Rev. Phys. Chem.* **13** (1962) 107; R. S. Mulliken, *J. Chim. Phys.* **61** (1964) 20.
[25] O. Hassel and C. Romming, *Quart. Rev. Chem. Soc.* **16** (1962) 1.
[26] L. Pauling, *Nature of the Chemical Bond*, Cornell Univ. Press, Ithaca, N.Y., 1944, p. 189.
[27] D. Hadzi, *The Hydrogen Bond*, Pergamon Press, New York, 1959.
[28] L. Hemholz and M. T. Rodgers, *J. Amer. Chem. Soc.* **61** (1939) 2590.
[29] T. C. Waddington, *Trans. Far. Soc.* **54** (1958) 25.
[30] G. Bessis, *Cahier de Phys.* **15** (1961) 105.
[31] E. Bauer and M. Magat, *J. Phys.* **9** (1938) 319.
[32] R. Grahm, *Arkiv för Fysik* **15** (1959) 257.
[33] S. Besnainou, S. Bratoz, and R. Prat, *J. Chim. Phys.* **61** (1964) 222.
[34] N. R. Sokolov, *Dokl. Akad. Nauk SSSR* **58** (1947) 611.
[35] C. A. Coulson and V. Danielson, *Arkiv. för Fysik* **8** (1954) 239.
[36] H. Margenau, *Rev. Mod. Phys.* **11** (1939) 1.
[37] F. London, *Trans. Far. Soc.* **33** (1937) 8.
[38] C. A. Coulson, *Industrie Chimique Belge*, No. 2 (1963) 149.
[39] J. Trotter, *Acta Cryst.* **11** (1958) 423.
[40] K. S. Pitzer and J. L. Hollenberg, *J. Am. Chem. Soc.* **75** (1953) 2219.
[41] M. S. Newman (ed.), *Steric Effects in Organic Chemistry*, Wiley, 1956, p. 57.
[42] E. B. Wilson, Tetrahedron **17** (1962) 191.
[43] J. B. Conn, G. B. Kistiakowsky, and E. A. Smith, *J. Am. Chem. Soc.* **61** (1939) 1868.
[44] L. S. Bartell, *J. Chem. Phys.* **32** (1960) 827.
[45] K. S. Pitzer (ed.), *Quantum Chemistry*, Prentice Hall, 1953, Section 9 and Appendix 8.
[46] J. B. Biot, *Mem. Acad. Sci. Toulouse* **2** (1817) 41.
[47] A. Cotton, *Comt. Rend. Acad. Soc.* **120** (1895) 989; *Ann. Chim. Phys.* **8** (1896) 347.

[48] R. de L. Kronig, *J. Opt. Soc. Am.*, **12** (1926) 547.
[49] H. A. Kramers, *Atti. Congr. Intern. Fisici* Como **2** (1927) 545.
[50] W. Moffitt and A. Moscowitz, *J. Chem. Phys.* **30** (1959) 648.
[51] C. Djerassi (ed.), *Optical Rotatory Dispersion*, McGraw-Hill, 1960, p. 164.
[52] E. V. Condon, W. Alton, and H. Eyring, *J. Chem. Phys.* **5** (1937) 753.
[53] W. Moffitt and A. Moskowitz, *J. Chem. Phys.* **30** (1939) 648.
[54] H. Eyring (ed.), J. Walter, and G. E. Kimball, *Quantum Chemistry*, Wiley, 1944, p. 346.
[55] C. Ouannes, *Thèse Science Paris*, 1964.
[56] T. L. Hill, *J. Chem. Phys.* **16** (1948) 399.
[57] A. L. H. Chung and M. J. S. Dewar, *J. Chem. Phys.*, in press.
[58] M. J. S. Dewar and N. L. Sabelli, *J. Phys. Chem.* **66** (1962) 2310;
[59] M. J. S. Dewar and A. L. H. Chung, *J. Chem Phys.* **39** (1963) 174;
[60] A. L. H. Chung, K. S. Dewar, and N. L. Sabelli, *Molecular Orbitals in Chemistry*, Acad. Press, New York.
[61] M. J. S. Dewar and H. N. Schmeising, *Tetrahedron* **11** (1960) 96.
[62] G. W. Wheland (ed.), *Resonance in Organic Chemistry*, Wiley, New York, 1955.
[63] E. Hückel, *Z. Phys.* **70** (1943) 204.
[64] L. M. Jackman, F. Sondheimer, Y. Amiel, D. A. Ben-Effrain, Y. Gaoni, R. Volovsky, and A. A. Bothmer-By, *J. Am. Chem. Soc.* **84** (1964) 4307.
[65] M. J. S. Dewar, *J. Am. Chem. Soc.* **74** (1952) 3353.
[66] M. J. S. Dewar and G. J. Gleicher, *J. Am. Chem. Soc.*, in press.
[67] H. C. Longuet-Higgins and L. Salem, *Proc. Roy. Soc.* A **251** (1959) 172.
[68] C. A. Coulson, *Conference on Quantum Mechanical Methods in Valence Theory*, Shelter Island, Long Island, N.Y. 1951, p. 42.
[69] F. J. Adrian, *J. Chem. Phys.* **28** (1958) 608.
[70] A. Almenningen and O. Bastiansen, *Kgl. Norske Videnskab Selskabs Skrifter*, No. 4 (1958).
[71] C. A. Coulson and A. Golebiewsky, *Tetrahedron* **11** (1960) 125.
[72] C. A. Coulson and D. Stockel, *Mol. Phys.* **2** (1959) 397.
[73] C. W. Haigh cited after C. A. Coulson, *Industrie Chimique Belge* **2** (1963) 149.
[74] J. B. Hendrickson, *J. Am. Chem. Soc.* **83** (1961) 4537.
[75] B. Hermann, *J. Org. Chemistry* **27** (1962) 441.
[76] C. F. Wilcox, S. Winstein, and W. G. McMillan, *J. Am. Chem. Soc.* **82** (1960) 5450.
[77] R. Pariser and R. G. Parr, *J. Chem. Phys.* **21** (1953) 767.
[78] J. Koutecky and J. Paldus, *Czechoslov. Chem. Commun.* **27** (1962) 599; J. Paldus, *Czechoslov. Chem. Commun.* **28** (1963) 1110, 2667.
[79] J. Paldus, *Czechoslov. Chem. Commun.* **27** (1962) 2139.
[80] See on this subject the important remarks by C. Dufraisse and Y. Lepage, *Compt. Rend. Acad. Sci.* **258** (1964) 5447.
[81] H. H. Jaffé and O. Chalvet, *J. Am. Chem. Soc.* **85** (1963) 156.

# CHAPTER III

# EQUILIBRIUM CONSTANTS OF REVERSIBLE REACTIONS IN SOLUTION

## 1. Introduction

We are now in a position to present the existing state of the quantum theory of balanced reactions in solution. We have just examined the part played by conformations in such reactions. Moreover we have demonstrated in Section 1 of Chapter II that, for a given conformation of the substances present, the equilibrium constant depends on six factors:

(a) the ratio of the distribution functions;

(b) the variation $\Delta\varepsilon_v$ of the vibration energy of the nuclei at absolute zero;

(c) the variation $\Delta\varepsilon_l$ of the energy of localised bonds;

(d) the variation $\Delta\varepsilon_d$ of the energy of delocalised bonds;

(e) the variation $\Delta\varepsilon_{n\cdot l}$ of the energy of interaction between unbonded atoms;

(f) and the variation $\Delta\varepsilon_s(T)$ of the solvation energy.

We now propose, on the basis of examples, to examine the relative importance of these different factors. We will concern ourselves mainly with examining how an equilibrium constant for a given type of reaction varies within a family of neighbouring molecules. In other words, we will concern ourselves first of all with the evolution of relative equilibrium constants. We will see that according to the different cases the variation in the family of any of the preceding factors one of the others may dominate and in this way become the determining cause of the behaviour of the reaction. In other more frequent cases, a number of factors will intervene simultaneously, frequently in opposite directions, so that the evolution of the constant depends on the difference between the contrary variations and becomes difficult to interpret.

## 2. The Equilibrium Between the Triplet State and the Singlet State of Biradicaloid Molecules

We know that molecules possessing an even number of electrons generally possess a singlet fundamental state. However, there are exceptions to this rule. Table 8-I of the second book (p. 36) recalls for example that the fundamental state of the oxygen molecule is triplet. We have also reserved special attention to molecules which thus present the triplet character in the fundamental state and which one calls biradical (second book, Section 29B). Between the biradicals and the usual molecules there is inserted a very interesting class of compound; that of the *biradicaloids*. One designates by this name molecules which, although singlets in the fundamental state, present a triplet excited state endowed with an energy which is hardly greater than that of the singlet state. Under these conditions this close triplet state is appreciably populated at ordinary temperature and for this reason the substance acquires *paramagnetic properties*. However, as the population of the triplet level depends on the temperature, *the paramagnetism considered varies with the temperature*. We are thus led to study the equilibrium constant $K$ between the molecules $M_T$ situated in the triplet state and the molecules which have remained in the fundamental state. We can represent this equilibrium by means of the formula

$$M_s \underset{K}{\rightleftharpoons} M_T$$

and according to the formula of Section 1, of Chapter II the constant $K$ is written

$$K = \frac{f_{Ms_T}}{f_{Ms_S}} \exp\left[ - \frac{\Delta\varepsilon_v + \Delta\varepsilon_l + \Delta\varepsilon_d + \Delta\varepsilon_{n\cdot l} + \Delta\varepsilon_s(T)}{\chi T} \right]$$

if the reaction takes place inside a solution.

We will restrict ourselves to following the evolution of the constant $K$ for one group of biradicaloids, namely that of the phenylquinodimethanes

Let $K$ and $K'$ be the constants relating respectively to two molecules of this series measured at the same time temperature and in the same solvent. One can then write

$$\frac{K'}{K} = \frac{f'}{f} \exp\left[ - \frac{\Delta\Delta\varepsilon_v + \Delta\Delta\varepsilon_l + \Delta\Delta\varepsilon_d + \Delta\Delta\varepsilon_{n\cdot l} + \Delta\Delta\varepsilon_s(T)}{\chi T} \right]$$

if one adopts the following symbols: the priority magnitudes relate to the second molecule, the values of $f$ represent the ratios of the partition functions and finally

$$\Delta\Delta\varepsilon_k = \Delta\varepsilon_k' - \Delta\varepsilon_k .$$

It is known that for a conjugated molecule of the type studied here the change in energy accompanying the passage from the fundamental state to a nearby excited state mainly affects the delocalised bond. Of the four first terms of the exponential it is therefore mainly $\Delta\varepsilon_d$ which risks dominating.

However, it is probable that the geometry of the molecule will be only slightly affected during this transition. Furthermore, Hückel's method shows that the distribution of the charges of the delocalised bond on the first triplet state remains the same as for the singlet state. The electrostatic interactions between the molecule and the solvent will therefore not depend very much on the state under consideration. One must therefore expect that the values of $\Delta\varepsilon_s(T)$ will remain low.

Finally, there is nothing to lead one to suppose that a ratio such as $f'/f$ will be far different from unity.

In all, one has the impression that it is $\Delta\varepsilon_d$ which will determine $K'/K$ and one is tempted to write

$$\frac{K'}{K} \# -\frac{\Delta\Delta\varepsilon_d}{\chi T}.$$

Table XIV summarizes [1] the values of $\Delta\varepsilon_d$ calculated by using Hückel's method and the experimental data relating to the magnetic properties of the first four terms of the family of tetraphenylquinodimethanes.

TABLE XIV

| Number $n$ of benzine rings | $\Delta\varepsilon_d$ (in $\beta$ units) | Nature of magnetic susceptibility | Value of this susceptibility for a 2% solution in benzine |
|---|---|---|---|
| 1 | 0.31 | diamagnetic | |
| 2 | 0.15 | diamagnetic | |
| 3 | 0.08 | paramagnetic | $\begin{cases} 26 \pm 4 & \text{at } 293\,°C \\ 82 \pm 2 & \text{at } 353\,°C \end{cases}$ |
| 4 | 0.04 | paramagnetic | $\begin{cases} 44 \pm 4 & \text{at } 293\,°C \\ 95 \pm 10 & \text{at } 353\,°C \end{cases}$ |

The appearance and then the increase of the paramagnetism as $n$ increases so that the population of the triplet level, and that is to say also the constant $K$, are increasing functions of $n$. Since, on the other hand, $\Delta\varepsilon_d$ decreases regularly when $n$ increases, it will be seen that, as expected, the *constant $K$ is a decreasing function of $\Delta\varepsilon_d$*.

## 3. Return to the Solvent Effect

Thus, in the example of equilibrium discussed during the analysis shown in Section 2 we observed that one factor seems to dominate all the others: the change of energy of the delocalised system $\Delta\varepsilon_d$. There are numerous cases where the situation is not so simple. The effect of the solvent frequently constitutes an important factor. Before dealing with particular cases, it seems essential to us to collect a few general data relating to this phenomenon.

Even when this solvent does not intervene as a true reagent, the solvation energy $\varepsilon_s$ results from a superimposition of elements of different natures.

One may divide this energy into four principal components [2]:

(a) the cavitation energy $\varepsilon_{sc}$ linked with the hole which the dissolved molecule produces in the solvent;

(b) the orientation energy $\varepsilon_{so}$ corresponding to the phenomena of partial orientation of the molecules of solvent caused by the vicinity of the solvated molecule;

(c) the isotropic interaction energy $\varepsilon_{si}$ corresponding mainly to the intermolecular forces with a long radius of activity;

(d) the anisotropic interaction energy $\varepsilon_{sa}$, being that resulting from the formation of the hydrogen bond at well localised points of the dissolved molecule.

One may therefore write

$$\varepsilon_s = \varepsilon_{sc} + \varepsilon_{so} + \varepsilon_{si} + \varepsilon_{sa}.$$

Here we will emphasize particularly the term $\varepsilon_{si}$* which in turn comprises at least three elements: the electrostatic energy $\varepsilon_{se}$, the polarisation energy $\varepsilon_{sp}$ and the dispersion energy $\varepsilon_{sd}$.

Let us first of all take the example of a solvent with a dielectric constant $D$. Born [3] has proposed a very simple means for estimating the electrostatic term $\varepsilon_{se}$.

This method consists in calculating the difference between the energy necessary for constituting the charge of the ion in a medium of dielectric constant $D$ and that corresponding to the same work carried out in vacuo. Therefore let $dq$ be an elementary charge which is brought from infinity to a small sphere of radius $r$ possessing a charge $q$ and immersed in a medium characterised by the constant $D$. If this charge $dq$ is located at a distance $x$ from the centre of the sphere, the elementary work $d^2\tau$ which corresponds to a displacement $dx$ may be written

$$d^2\tau = \frac{q\,dq}{Dx^2}\,dx.$$

The total work corresponding to the contribution of $dq$ will therefore be

$$d\tau = \frac{q\,dq}{D}\int_{\infty}^{r}\frac{dx}{x^2} = -\frac{q\,dq}{Dr}.$$

Then let $Ze$ be the charge of the ion, when the energy associated with the formation of this charge can be written

$$\tau = -\int_{0}^{Ze}\frac{q\,dq}{Dr} = -\frac{Z^2e^2}{2Dr}.$$

From this one immediately gets

$$\varepsilon_{se} = -\frac{Z^2e^2}{2r}\left(1 - \frac{1}{D}\right).$$

The effective application of this formula requires a knowledge of the radius $r$. Latimer, Pitzer and Slansky [4] have proposed an empirical process for choosing these radii. This process leads to the adoption of crystalline ionic radii increased by 0.1 Å in the case of anions and by 0.85 Å in the case of cations when water is used as

---

* The very profusely documentated lecture given by Professor Souchay in 1965 to the Colloque National organised by Professor Josien at Bordeaux has helped us considerably in dealing with this subject here.

the solvent. This difference doubtless arises from the fact that the organisation of the molecules of water is more compact in the vicinity of an anion than in the vicinity of a cation because in the former case it is the hydrogen atoms of the water molecules which tend to direct themselves towards the ion.

Noyes [5], observing that the 'effective' radii thus calculated lead to too high electrostatic entropies of solvation prefers to come back to the use of ionic radii and to introduce an effective dielectric constant $D_e$ into Born's formula.

Apart from this, it is natural to think that in fact one has to introduce such a constant $D_e$. More precisely, the partial orientation of the molecules of solvent in the vicinity of the ion brings about a modification in the mean specific inductive power of the solvent near this ion.

Glueckhauf [6] took up once again the calculation of the work by taking into account the variation in the dielectric constant with $x$.

Stokes [7] reconsidered the whole of the problem in the case of water. He thinks that when an ion is solvated it is natural to adopt its ionic radius for the value of $r$ because the entropies of solution of salts are usually low. He thinks, on the other hand, that in order to evaluate the charge work in vacuo it is best to choose the Van der Waals' radius $R$. He also assumes that for an anion one may use the normal constant $D$ (that is 78 in the case of water). One then obtains

$$\varepsilon_{se} = -\left(\frac{Z^2 e^2}{2R} - \frac{Z^2 e^2}{2Dr}\right).$$

For a cation, he observes that in a monomolecular or bimolecular layer of water surrounding the ion (according to whether it is monovalent or divalent ($Z=1$ or 2)) the dielectric constant is lowered to $D_e$ and it rapidly takes back its normal value beyond this layer.

If one adopts the thickness 2.8 for a monomolecular layer of water, one is then led to write

$$d\tau = +\frac{q\,dq}{D}\int\limits_{\infty}^{r+2.8Z}\frac{dx}{x^2} + \frac{q\,dq}{D_e}\int\limits_{r+2.8Z}^{r}\frac{dx}{x^2} =$$

$$= -q\,dq\left\{\frac{1}{(r+2.8Z)\,D} + \frac{2.8Z}{r\,(r+2.8Z)\,D_e}\right\}$$

from which one finally gets

$$\varepsilon_{se} = -\frac{Z^2 e^2}{2}\left\{\frac{1}{R} - \frac{1}{(r+2.8Z)\,D} - \frac{2.8Z}{r\,(r+2.8Z)\,D_e}\right\}.$$

In practice Stokes is led to choose

$$D_e = 9$$

Let us now suppose that we wish to evaluate $\varepsilon_{se}$ for a molecule or a multipolar ion such as for example a heteroatomic conjugated molecule.

Considering each atom as a spherical ion with an apparent charge $Q_i$ and generalising the formula of Born [8] one immediately reaches the expression

$$\varepsilon_{se} = -\left(\sum_i \frac{Q_i^2}{2r_i} + \sum_{i<j} \frac{Q_i Q_j}{r_{ij}}\right)\left(1 - \frac{1}{D}\right)$$

if $r_i$ indicates the ionic radius of the atom $i$ and $r_{ij}$ signifies the distance between the atom $i$ and the atom $j$. One can in any case subject this formula to modifications analogous to those which have been described in the case of a spherical ion, like those connected with the introduction of the effective radii of Latimer, Pitzer and Slansky or the consideration of an effective dielectric constant according to Stokes' process.

We think that it is as well to point out that apart from Born's method and its different extensions there exist more structural methods which should in principle call for our special attention. In actual fact, since their present form does not yet permit of their use in the case of molecules which are rather complex, we will restrict ourselves to referring to them very briefly.

Eley and Evans [9] have proposed a very simple way for evaluating the entropy of solvation of a spherical ion in water.

One evaluates first of all the cavitation energy $\varepsilon_I$ corresponding to the transfer to the gaseous phase of a tetrahedral group of five molecules of water. One then considers the dissociation energy $\varepsilon_{II}$ of the tetrahedron of five isolated molecules of water. To estimate this energy one only takes into account the electrostatic interactions between the central molecule of water and the molecules of water surrounding it. Of these four molecules one assumes that two present their hydrogen to the oxygen of the central molecule and that the other two present their oxygen to the hydrogen atoms of the latter. One calculates the electrostatic energy of an interaction of the central molecule with its neighbours by attributing to the atoms charges which are susceptible of reproducing the polar moment of a molecule of water.

In a third stage one faces four molecules of water around the ion whose solvation it is intended to study and one calculates the energy $\varepsilon_{III}$ corresponding to the electrostatic interactions between the ion of charge $Ze$ and the charges $Q_H$ and $Q_O$ attributed to the atoms of water. One therefore has

$$\varepsilon_{III} = \sum_i \frac{ZeQ_i}{r_i}$$

if $r_i$ designates the distance of the atoms of hydrogen or oxygen of charges $Q_i$ to the centre of the ion. One finds that this energy is stronger for an anion than for a cation because the distances $r_i$ are relatively less in the former case.

The fourth operation consists in introducing the ion thus solvated into the cavity which one assumes possesses suitable dimensions. The energy of interaction between the solvated ion and the rest of the liquid is divided into two parts:

(a) the energy of interaction between the solvated ion and the molecules of water immediately adjacent $\varepsilon_{IV\alpha}$;

(b) the energy connected with the reorganisation of the molecules of water in the rest of the solvent $\varepsilon_{IV\beta}$.

Eley and Evans have estimated this latter term by using the approximation of Born, which gives

$$\varepsilon_{IV\beta} = -\frac{Z^2 e^2}{2\,(r + 2.8)} \left(1 - \frac{1}{D}\right).$$

The fifth and last stage of the calculation is bound up with the necessity of bringing into the liquid phase the last molecule of water which is still in the gaseous phase, hence an energy $\varepsilon_V$.

The enthalpy of hydration is written

$$\Delta H_h = \varepsilon_I + \varepsilon_{II} + \varepsilon_{III} + \varepsilon_{IV\alpha} + \varepsilon_{IV\beta} + \varepsilon_V.$$

One finds finally

$$\Delta H_h = 19 - E_{III} - E_{IV\beta} \quad \text{for an anion}$$
$$\Delta H_h = 31 - E_{III} - E_{IV\beta} \quad \text{for a cation}$$

provided that these entities are solvated by four molecules of water. Table XV contains a few numerical results of the terms $\varepsilon_{III}$ and $\varepsilon_{IV\beta}$ expressed in kcal mole$^{-1}$.

TABLE XV

|  | F$^-$ | K$^+$ | Al$^{+++}$ |
|---|---|---|---|
| $\varepsilon_{III}$ | 82.4 | 69.6 | 569 |
| $\varepsilon_{IV\beta}$ | 39.8 | 39.8 | 441 |

It is interesting to note that the energy bound up with the action of the solvated ion on the solvent is never negligible and becomes particularly large in the case of an ion possessing a powerful charge.

Various authors have improved the theory of Eley and Evans, particularly Buckingham, who has taken into account a quadrupolar moment, the energy of polarisation and the energy of dispersion.

## 4. Polarographic Semi-Wave Potential of Conjugated Molecules

A. THE CASE OF HYDROCARBONS

Laitinen and Wawzonek [10] observed as far back as 1942 the possibility of reducing conjugated molecules on the dropping mercury electrode. They assumed that the reaction took place according to the following mechanism

$$\begin{aligned}
R + e^- &\rightleftharpoons R^- \\
R^- + e^- &\rightarrow R^{-2} \\
R^{-2} + 2H_2O &\rightarrow RH_2 + 2OH^-
\end{aligned}$$

and they thought that the polarographic wave which they were observing corresponded to the first stage of this reduction. Not long afterwards [11], however, they observed that in a 75% dioxane solution using tetrabutyl ammonium iodide as the indifferent electrolyte, the first stage of the reaction seemed to correspond to the fixation of two electrons for certain molecules such as phenanthrene.

For his part, Bergman [12] observed that in monomethyl ethyleneglycol ether in the presence of the same electrolyte the first polarographic wave always corresponds to the fixation of a single electron.

Hoijtink et al. [13] examined the effect of the pH on the phenomenon and showed that in the case of dioxane one easily passes for one and the same hydrocarbon to a polarographic wave with two steps showing the successive addition of two electrons to a wave with a single step suggesting the simultaneous fixation of the two electrons when one increases the acidity of the medium.

The authors therefore assume that in the presence of protons $H^+$ the reactions take place according to the scheme

$$R + e^- \rightleftharpoons R^- \tag{1}$$
$$R^- + H^+ \rightarrow RH \tag{2}$$
$$RH + e \rightarrow RH^-. \tag{3}$$

In the presence of an appreciable quantity of protons stages (2) and (3) would become very rapid and would prevent the showing of the wave corresponding to stage (1).

As we are concerned here mainly with studying the equilibrium

$$R + e^- \rightleftharpoons R^-$$

we would therefore attach more importance to measurements carried out in an aprotic solvent such as those of Bergman.

To understand how it is possible to link up the result of such measurements and the quantum magnitudes it will be as well for us first of all to establish the equation which connects the polarographic semi-wave potential with the constant $K$ of an oxidation-reduction equilibrium.

Let us therefore consider such an equilibrium

$$A + B + \cdots \overset{K}{\rightleftharpoons} C + D + \cdots.$$

Let us suppose that A represents the oxidised form R, and C represents the reduced form $R^-$.

Let us assume that we are constructing the polarogram, that is to say that we are representing the variation of the current intensity as a function of the potential difference $E$ applied between the mercury electrode and the reference electrode. If we operate in the presence of an appreciable concentration of indifferent electrolyte we will observe the following phenomena. Below the potential necessary for reduction, one observes the existence of a weak current which is called the residual current. As soon as the reduction potential is reached, the current increases. If one continues to raise $E$ gradually the current continues to increase and then reaches a saturation value. The polarogram then has the shape of an S.

At this moment all the molecules A touching the electrode are reduced and the current to some extent measures the speed of diffusion of the molecules A towards the drops of mercury. However, since the potential corresponding to this current is difficult to measure with accuracy, it is preferred to characterise the phenomenon by the potential corresponding to the half current which is known as the polarographic semi-wave potential $E_{1/2}$.

Let $D_A$ and $D_C$ be the diffusion coefficients of the molecules A and C, and one can then show [14] that one then has

$$\frac{[C]}{[A]} = \sqrt{\frac{D_A}{D_C}}.$$

Furthermore it is known that the potential of a reversible reaction is connected to the variation of free energy $\Delta G$ by the relationship

$$E = -\frac{\Delta G}{nF}$$

if F denotes the Faraday and $n$ the number of electrons necessary for reduction. If $\Delta G^0$ denotes the variation in energy at equilibrium

$$\Delta G = \Delta G^0 + RTL\frac{(C)(D)\dots}{(A)(B)\dots}$$

the brackets denoting the activities.

Therefore

$$E = E^0 - \frac{RT}{nF}L\frac{(C)(D)}{(A)(B)}$$

but

$$E^0 = -\frac{\Delta G^0}{nF} = +\frac{RT}{nF}LK + \frac{RT}{nF}L\gamma_{eq}$$

if $\gamma_{eq}$ denotes the ratio of the activity coefficients at equilibrium

$$\gamma_{eq} = \frac{\gamma_C\gamma_D\cdots}{\gamma_A\gamma_B\cdots}.$$

Finally we get

$$E = \frac{RT}{nF}\left(LK + L\gamma_{eq} - L\frac{[C][D]\dots}{[A][B]\dots} - L\gamma\right)$$

if $\gamma$ denotes the ratio of the activity coefficients outside equilibrium. From this one immediately gets

$$E_{1/2} = \frac{RT}{nF}\left(LK + L\gamma_{eq} - \tfrac{1}{2}L\frac{D_A}{D_C} - L\frac{[D]\dots}{[B]\dots} - L\gamma\right).$$

Thus one has established the link between $E_{1/2}$ and $K$ and if one introduces into this relation the general expression of $K$, one immediately obtains

$$E_{1/2} = \frac{1}{n\mathrm{F}} \left( RTLf - \Delta\varepsilon_v - \Delta\varepsilon_l - \Delta\varepsilon_d - \Delta\varepsilon_{n\cdot l} - \Delta\varepsilon_s(T) + \right.$$
$$\left. + RTL\frac{\gamma_{eq}}{\gamma} - \tfrac{1}{2}L\frac{D_A}{D_C} - L\frac{D\dots}{B\dots} \right).$$

Or again

$$nE_{1/2} = -\Delta\varepsilon_v - \Delta\varepsilon_l - \Delta\varepsilon_d - \Delta\varepsilon_{n\cdot l} - \Delta\varepsilon_s(T) + $$
$$+ \chi T\left( Lf + L\frac{\gamma_{eq}}{\gamma} - \tfrac{1}{2}L\frac{D_A}{D_C} - L\frac{D\dots}{B\dots} \right)$$

if the energies are expressed in electron-volts.

Let us now suppose that we choose the calomel electrode as the reference electrode. The reaction may be written

$$\mathrm{R} + e\,(\mathrm{Hg}) \rightleftharpoons \mathrm{R}^-$$
$$\underline{\mathrm{Hg}\,(\mathrm{liquid}) + \mathrm{Cl}^- \rightleftharpoons \mathrm{HgCl}\,(\mathrm{crist}) + e\,(\mathrm{Hg})}$$
$$\mathrm{R} + \mathrm{Hg}\,(\mathrm{liquid}) + \mathrm{Cl}^- \rightleftharpoons \mathrm{R}^- + \mathrm{HgCl}\,(\mathrm{crist})$$

and finally

$$E_{1/2} = -\Delta\varepsilon_v - \Delta\varepsilon_l - \Delta\varepsilon_d - \Delta\varepsilon_{n\cdot l} - \Delta\varepsilon_s(T) + $$
$$+ \chi T\left( Lf + L\frac{\gamma_{eq}}{\gamma} - \tfrac{1}{2}L\frac{D_R}{D_{R^-}} - L\frac{[\mathrm{HgCl}]_{\mathrm{cryst}}}{[\mathrm{Hg}]_{\mathrm{liq}}\,[\mathrm{Cl}^-]} \right).$$

It is easy to guess which will be the most important of these nine terms when one tries to follow the evolution of $E_{1/2}$ in a series of neighbouring hydrocarbons. There is not much reason to think that the coefficient of diffusion $D_{R^-}$ will be very different from $D_R$ and still less that $D_R/D_{R^-}$ will vary a great deal from one hydrocarbon to another.

One may quite reasonably ignore all the terms in the brackets.

Since the electron which is fixed onto R during the course of the reduction increases by one unit the number of electrons of the delocalised bond, it is natural to think that the variation of $\Delta\varepsilon_d$ will dominate that of $\Delta\varepsilon_v$, $\Delta\varepsilon_l$ or $\Delta\varepsilon_{n\cdot l}$. The variation of the term $\Delta\varepsilon_s(T)$ connected with the solvation energy also runs the risk of being considerable since one passes from a neutral molecule to an ion during the reduction. One may therefore assume that as a first approximation for a series of neighbouring hydrocarbons

$$E_{1/2} = -\Delta\varepsilon_d - \Delta\varepsilon_s(T) + \mathrm{Ct}$$
$$= \quad \mathrm{A} - \Delta\varepsilon_s(T) + \mathrm{Ct}'$$

since $\Delta\varepsilon_d$ here represents the essential of what it has been agreed to call the electronic affinity of the hydrocarbon.

It was Maccoll [15] who first noted the existence of a linear relationship between $E_{1/2}$ and the electronic affinity. To represent the electronic affinities he used the energies associated with the first virtual orbital calculated by Hückel's method. He also based himself on the measurements of Wawzonek. Pullman *et al.* [16] stated that if one considers a large range of conjugated hydrocarbons one does in fact obtain several straight lines even if one adjusts the resonance integral $\beta$ to the different lengths of the bonds of the molecules. Hoijtink and Van Schooten [17] have shown that if one uses $m\beta$ to denote the resonance energy of the virtual orbital one has more or less

$$E_{1/2} = (m\beta - 1.05) \, \text{V} \, .$$

Hoijtink [18] continued the discussion of the problem, taking Bergman's measurements into account.

He showed that, whatever may be the solvent, one finally has rather

$$E_{1/2} = (m\beta - 0.85) \, \text{V}$$

whether one uses Hückel's method or Wheland's (second book, p. 188) to calculate $m$. The value which should be given to $\beta$ is of the order of $-2.55$ for Hückel's method and $-2.25$ for Wheland's method and does not depend a great deal on the solvent (75% or 95% dioxane, or monomethyl ethyleneglycol).

Hoijtink also examined the reaction

$$\text{R}^- + e \rightleftharpoons \text{R}^{-2}$$

and found that

$$E_{1/2} = (m\beta - 1.05) \, \text{V} \, .$$

He also noted that the case of the diarylpolyenes, however, constitutes a slight exception to these rules, the relationship between $E_{1/2}$ and $m$ no longer being very linear. He thinks that this fact is connected with another mechanism being responsible for the reduction. However, it is necessary to note that, as we have seen, the delocalisation is not great in a polyene chain (Section 3 of Chapter II).

Therefore without doubt Hückel's and Wheland's methods provide fairly poor values of $m$.

We will also mention that Hoijtink has studied the semi-wave potential of non-plane hydrocarbons. He has observed, for example, that $E_{1/2}$ of rubrene is more or less the same as that of naphthacene, which seems to point out that in rubrene the groups are not conjugated with the naphthacene radical probably because of the quasi-perpendicularity of the planes of these groups.

Let us come back to the theoretical relationship

$$E_{1/2} = A - \Delta\varepsilon_s(T) + \text{Ct} \, .$$

From this, with Matsen *et al.* [19, 20] one may deduce that

$$\Delta\varepsilon_s(T) + \text{Ct} = E_{1/2} - A \, .$$

It is known that these authors (second book, p. 210) estimated the electronic affinities by identifying them with the energy of the first virtual orbital calculated by the method of Pariser and Parr.

Thus they observed that $\Delta\varepsilon_s(T)+Ct$ varies from 3.8 eV for benzene to 2.6 eV for anthracene, passing through 3 eV for naphthalene. In other words, the variation of $\Delta\varepsilon_s(T)$ is not negligible.

Jano [21] continued the study of this factor. He first of all tried to estimate as carefully as possible the electronic affinities $A$. In order to do this he first of all improved the calculation of the ionisation energies $I$ by using the approximation of Pariser and Parr but taking into account the charge effects which accompany the departure of the electron according to a process similar to that followed by Hoyland and Goodman [22].

Jano observed that, even when calculated in this way, the sum of the ionisation energy and the electronic affinity is more or less constant.

If furthermore one wishes to get the few values of $A$ measured by Becker and Wentworth [23, 24] it is necessary to write

$$I + A = 8.2 \text{ eV}.$$

Jano finally obtained the $A$ of this relationship from the values of $I$ calculated according to the above-mentioned process.

When confirming the results of Pullman *et al.* he observed that one obtains several straight lines (Figure 18) when one plots $E_{1/2}$ against A for a large series of aromatic hydrocarbons. The magnitude $E_{1/2}-A$ thus measures $\Delta\varepsilon_s(T)$ apart from a constant.

Jano also calculated $\Delta\varepsilon_s(T)$ from Born's formula as generalised by Hoijtink *et al.* He obtained a remarkable linear relationship between the values of $E_{1/2}-A$ and those of $\Delta\varepsilon_s(T)$ calculated in this way (Figure 19).

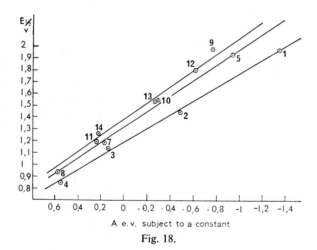

Fig. 18.

These facts seem to demonstrate:

(a) that Born's generalised formula is well suited for the estimation of $\Delta\varepsilon_s(T)$;

(b) that the origin of the plurality of the straight lines expressing the relationship between $E_{1/2}$ and A is bound up in the main with the presence of the term $\Delta\varepsilon_s(T)$.

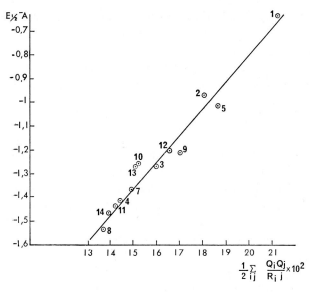

Fig. 19.

Relationships between the polarographic semi-wave potentials and various other magnitudes have been observed. Let us consider once again, for example, Hoijtink's relationship

$$E_{1/2} = (m\beta - 0.85)\ \text{V}.$$

It is known that for an alternant hydrocarbon (second book, section 29) the quantity $2m\beta$ provides an approximation of the transition energy of the band $p$ within the framework of Hückel's method.

One must therefore expect to observe a linear relationship between $E_{1/2}$ and this transition energy.

In point of fact one observes relationships of this type for different families of conjugated hydrocarbons [25, 26].

These relationships do not seem to us to be fundamental. They are only verified insofar as the energy of the band $p$ is linked with the electronic affinity. In cases where this connection disappears (in the case of rubrene for example) the relationship between $E_{1/2}$ and the band $p$ is falsified.

Basu and Bhattacharya [27] also observed the existence of a relationship between $E_{1/2}$ and the lowest energy of paralocalisation* in the case of alternant hydrocarbons.

Fernandez and Domingo [28] have confirmed the existence of this relationship.

Given [29] has also pointed out that one may also show a relationship between $E_{1/2}$ and the greatest energy of localisation of the alternant hydrocarbons** but

* The definition of this magnitude is to be found in Section 18 of the first book.
** The definition of this idea will be found in Section 3, Chapter IV of this book.

that all these relationships with the energies of localisation probably do not present a fundamental character.

Lund [30] has shown that it is possible to *oxidise* conjugated hydrocarbons electrochemically by means of a rotating platinum electrode. The phenomenon seems to involve the transfer of two electrons from the hydrocarbon to the electrode

$$R \rightleftharpoons R^+ + e^-$$
$$R^+ \rightleftharpoons R^{++} + e^-.$$

Hoijtink [31] reported that one should therefore observe a relationship between the semi-wave potential corresponding to this phenomenon and the energy associated with the highest orbital occupied by the hydrocarbon which, as one knows, is connected with its ionisation energy. He has shown that such a relationship even exists in the case of non-alternant hydrocarbons.

In the case of alternant hydrocarbons it is known that if

$$\alpha + m\beta$$

denotes the energy associated with the highest orbital occupied, the energy associated with the first virtual orbital is

$$\alpha - m\beta$$

(second book, p. 170).

Hoijtink deduces from this that the $E_{1/2}$ of oxidation must in this case be connected by a linear relationship with the $E_{1/2}$ of reduction. This is in fact what he observes. He also reports that, as expected, this latter relationship ceases to hold good for the non-alternant hydrocarbons.

B. THE CASE OF OTHER CONJUGATED MOLECULES

Aalbersberg and Mackor [32] studied the polarographical reduction of alternant hydrocarbons such as perylene, tetracene, anthracene in a very acid medium. This medium contained boron trifluoride and trifluoracetic acid. They assume that the first polarographical wave corresponds to the oxidation of the positive ion of the hydrocarbon according to the equilibrium

$$R^+ + e^- \rightleftharpoons R^*$$

which would be followed by an irreversible protonation

$$R + H^+ \rightarrow RH^+$$

The potentials $E_{1/2}$ corresponding to this wave vary, in fact, like those measured by Lund.

The second wave which is observed seems to be bound up with the oxidation of the

---

* $R^+$ would arise from the oxidation of $RH^+$ by the oxygen of the air.

$RH^+$ ion formed according to

$$RH^+ + e \quad \rightleftharpoons RH$$
$$RH + H^+ \rightarrow RH_2^+$$
$$RH_2^+ + e^- \rightarrow RH_2$$

The corresponding semi-wave potential must therefore be connected with the electronic affinity of the ion $RH^+$.

The structure of such ions is now well known (see Section 6). In the case of anthracene. for example, it is known that we are dealing with the following ion

It contains an alternant conjugated system with an odd number of atoms. The highest orbital occupied is therefore used only once in the representation of the fundamental state of the delocalised bond and in the framework of Hückel's method the energy associated with this orbital is therefore simply equal to the integral $\alpha$ (second book, p. 170). The electronic affinity of the ion is therefore itself equal to $\alpha$. Since this line of reasoning can be generalised to all the $RH^+$ ions derived from alternant hydrocarbons, one should expect that all the $E_{1/2}$ should be equal. In point of fact this is not the case. They vary between 0.7 and 1.2 eV. Aalbersberg and Mackor thought that this disagreement between the experimental data and the results of calculation arose from an insufficiency of the approximation of Hückel. They therefore calculated the electronic affinities of the $RH^+$ using the method of Hush and Pople [33] based on the approximations of Pariser and Parr (second book, Section 36). They obtained a linear relationship between the affinities thus calculated and the potentials $E_{1/2}$ under consideration.

The methylated derivatives of the conjugated hydrocarbons have been studied by various authors [34]. There again $E_{1/2}$ varies more or less like the calculated electronic affinity. However, the agreement is not very satisfactory in the case of azulene.

Anthoine et al. [35] measured the polarographic reduction semi-wave potentials of monoaza-aromatic derivatives in anhydrous dimethylformamide in the presence of tetraethyl ammonium. They obtained the relationship

$$- E_{1/2} = (- m\beta + 0.484) \text{ V}$$

in which

$$\beta = - 2.12 \text{ eV}.$$

The case of nitrated derivatives has been examined by Basu and Chandhuri [36] and by Zahradnik and Bocek [37]; Schmid and Heilbronner [38] have studied that of aldehydes such as benzaldehyde and those deriving from naphthalene and anthracene.

Klopman and Nasielski [39] obtained the relationship

$$- E_{1/2} = (- m\beta + 0.93) \, V$$

in which $\beta = -2.42$ eV in the case of ethyl esters derived from polycyclic aromatic acids. Finally we will refer to studies of the same type relating to the derivatives of furane [40] and xanthone [41].

## 5. Other Oxidation-Reduction Equilibria

### A. REDUCTION POTENTIAL OF CONJUGATED HYDROCARBONS

Having observed that biphenyl is one of the conjugated hydrocarbons which possesses the weakest electro-affinity, Hoijtink et al. [42] had the idea of studying the equilibrium

$$biphenyl^- + R \rightleftharpoons biphenyl + R^-$$

in which R represents another conjugated hydrocarbon.

They were thus led to carrying out the potentiometric titration of a solution of conjugated hydrocarbons in dimethoxyethane or tetrahydroxyfurane by a solution of biphenyl sodium in the same solvent.

A simple reasoning would show us that the reduction potential thus measured for R must depend in the main on the difference between the energy of the delocalised bonds of the entities of the second member of the corresponding energy for the first member of the equilibrium under consideration. In other words, we must expect to observe a linear relationship between these potentials and the difference of electro-affinity between R and biphenyl.

Hoijtink et al. have in fact observed that the potential $E$ obeys the equation

$$E = (m\beta + 1.95) \, V$$

in which

$$\beta = - 2.19 \, eV .$$

### B. OXIDATION POTENTIAL OF PHENOLS

Fueno et al. have studied in a similar manner the oxidation potentials of numerous phenols [43].

They assume that the stage which determines the potential is the transfer of an electron from the phenol to the oxidising agent

$$Ox + ROH \rightleftharpoons Ox^- + ROH^+ .$$

The reaction would continue in the loss of a proton

$$ROH^+ \rightarrow RO + H^+ .$$

They therefore looked for and found a linear relationship between the oxidation potential of the phenols and the energy associated with the highest orbital occupied.

## C. OXIDATION-REDUCTION POTENTIAL OF QUINONES

It is usually assumed [44] that the reactions which determine the oxidation-reduction potentials of quinones may be written

One may therefore expect that one of the important terms is the difference $\Delta \varepsilon_d$ between the energy of the delocalised system of hydroquinone and that of the corresponding quinone. Numerous authors have calculated this term, using more and more elaborate methods.

In the first work one went as far as supposing that the quinone groups did not belong to the delocalised system and that the effect of the OH group on the delocalised system of hydroquinones was negligible. It was then assumed that the delocalised bond of a quinone [45–50] possessed the same energy as that of the corresponding quinodimethane.

More recently, Evans *et al.* [51, 52] Gold [53] and Deschamps [54] have calculated $\Delta \varepsilon_d$ by the method of Pauling and Wheland (second book, p. 225). Finally Le Bihan [55] has used the method of the auto-coherent field in the approximation of Pariser and Parr.

In all the cases one observes a good linear relationship between $\Delta \varepsilon_d$ and the oxidation-reduction potential for the paraquinones and another parallel relationship for the orthoquinones. The existence of these two relationships is not surprising. It may be explained by assuming that the difference between $\Delta \varepsilon_l$ for a paraquinone and $\Delta \varepsilon_l$ for an orthoquinone is more or less constant, which would seem normal if one were to suppose that the essential of this difference arises from the hydrogen bond which occurs between the two OH groups of an orthohydroquinone.

We will add that Le Bihan (*loc. cit.*) observed that the part played by $\Delta \varepsilon_s(T)$ is negligible here, unlike what happens in the case of the reduction of conjugated hydro-carbons. This fact is explained if one observes that in the latter case one passes from a neutral molecule to an ion, whereas in the oxidation-reduction equilibrium of quinones one compares energies relating to two molecules which are globally electrically neutral.

Let us observe finally that A. Pullman [56] noted that the oxidation-reduction potential of the quinones:

(a) increases when the lowest free orbital of the quinone falls;
(b) falls with the highest orbital occupied of the hydroquinone.

## D. BIOCHEMICAL APPLICATIONS

Assuming that this observation may be extended to numerous reactions, one can understand better the conclusions presented in Section 44 of the second book and in particular why ferrocytochrome C is a powerful reducing agent whilst ferricytochrome C acts as an active oxidising agent.

A. Pullman [57], however, has pointed out that such an extension may be wrong. Mrs Pullman has, in fact, studied the origin of the relationship which exists in the case of the quinones between the oxidation-reduction potentials and the energies of the orbitals. This study shows clearly that one must not expect this relationship to be a general one. A. Pullman has even shown that in the case of diphosphopyridine nucleotide and its analogues there is no correlation between the potential and the orbitals concerned in the electron transfer.

In any case it is easy to understand why. Various authors [58]–[60] have just shown that the mechanism of operation of D.P.N. as an oxidation-reduction co-enzyme comprises the direct transfer of an $H^-$ ion from the substratum to position 4 of the oxidised form of this enzyme

It is therefore the nucleophilic localisation energy of position 4 which should intervene in the determination of the potential. The calculations of A. Pullman confirm this idea, whilst the works of Kaplan et al. [61, 62] showed the existence of a parallelism as between the oxidation-reduction potential of the analogues of D.P.N. and the aptitude of their apex 4 for fixing a negative ion.

## 6. Strength of Acids and Bases

### A. INTRODUCTION

Let us consider an acid-base equilibrium

$$AH \underset{K_a}{\rightleftharpoons} A^- + H^+ .$$

The constant $K_a$ which by definition measures the strength of the acid may be expressed according to the equation

$$K_a = \frac{f_A \cdot f_{H^+}}{f_{AH}} \exp\left[ - \frac{\Delta\varepsilon_v + \Delta\varepsilon_l + \Delta\varepsilon_d + \Delta\varepsilon_{n \cdot l} + \Delta\varepsilon_s(T)}{\chi T} \right].$$

From this one gets

$$M p_{K_a} = L \frac{f_{AH}}{f_{A^-} f_{H^+}} + \frac{\Delta\varepsilon_v + \Delta\varepsilon_l + \Delta\varepsilon_d + \Delta\varepsilon_{n \cdot l} + \Delta\varepsilon_s(T)}{\chi T}$$

if $M$ represents the factor necessary for the transformation of the vulgar logarithm into a Napierian logarithm.

## B. STRENGTH OF AMINO-ACIDS

Without trying to follow the chronological development of the analysis of the idea of $p_K$ by means of quantum theories, we will first of all examine a few studies connected

$$R \qquad R'$$
$$\diagdown \diagup$$

with the strength of amino-acids $H_2N$————$C$————$CO_2H$, the formulae of which have been set out in Section 43 of the second book.

For such substances it is necessary to consider two $p_K$'s. The $p_{K1}$ corresponds to the

$$R \qquad R'$$
$$\diagdown \diagup$$

passage from the cationic form $H_3^+N$————$C$————$CO_2H$ to the dipolar form

$$R \qquad R'$$
$$\diagdown \diagup$$

$H_3^+N$————$C$————$CO_3^-$ and $p_{K2}$ corresponds to the appearance of the anionic form

$$R \qquad R'$$
$$\diagdown \diagup$$

$H_2N$————$C$————$CO_2^-$.

It is clear that in this case the term $\Delta \varepsilon_d$ will no longer necessarily be the most important. On the contrary, one feels that $\Delta \varepsilon_l$ will be essential if it is a question of comparing the anionic form with the dipolar form. We are thus led to set out precisely the method which can be used for calculating the energy of the localised bonds of a molecule such as an amino-acid.

The method of Del Ré [63] doubtless constitutes one of the most convenient processes. This process is connected with the method of Sandorfy and Daudel (Section 18C, second book) in that it deals with the extension of the method of Pauling and Wheland to the case of saturated molecules.

In Del Ré's method the orbital $Y$ associated with a localised bond is expressed in the form of a linear combination of the two corresponding atomic hybrids $\Psi_A$ and $\Psi_B$

$$Y = a\psi_A + b\psi_B.$$

The coefficients $a$ and $b$ and the energy $\varepsilon$ associated with the orbital $Y$ are then calculated in the approximation of Pauling and Wheland as a function of the parameters $\alpha_A$, $\alpha_B$ and $\beta_{AB}$.

The total energy of the delocalised bonds is obtained by calculating the sum of the values of $\varepsilon$ thus obtained for the different bonds. Each orbital of the bond $Y$ is therefore treated as if it were alone. The influence of the other bonds on one of them is only introduced by suitably modifying the parameters $\alpha_A$ and $\alpha_B$, thus taking the inductive effects into account. The entire skill of this process resides in the choice of parameters.

More precisely, if one supposes

$$\alpha_A = \alpha + \xi_A \beta$$

one will calculate $\xi_A$ by means of the formula

$$\xi_A = \xi_{A^0} + \sum_P \zeta_{AP}\xi_P$$

the sum being extended to all the atoms attached to A, $\xi_{AO}$ characterising the isolated atom A and the values of $\zeta_{AP}$ characterising the different bonds. Del Ré (*loc. cit.*) has proposed adjusted parameters so as to take into account the polar moments of the different bonds.

Del Ré et al. [64, 65] have examined the structure of the amino-acids and calculated the principal terms, determining their values of $p_K$ by means of this very simple method. When necessary, $\Delta\varepsilon_d$ is calculated by means of the classic method of Pauling and Wheland.

Since these methods do not explicitly introduce the electrostatic interactions appearing between the total charges carried by the atoms, the authors add to $\Delta\varepsilon_l + \Delta\varepsilon_d$ a term of the form

$$\Delta \sum_{\mu < \nu} Q_\mu Q_\nu / r_{\mu\nu}$$

in which $Q$ signifies these charges and $r$ signifies the inter-atomic distances. This term is particularly important because of the fact that during the protonation a new charged atom intervenes.

Del Ré et al. have obtained in this way fairly satisfactory linear relationships between the values of $p_{K1}$ and the sum of the three energy terms under consideration. In the case of $p_{K2}$ the relationship is less satisfactory. We will also observe that the authors noted that $\Delta\varepsilon_s(T)$ does not vary much from one amino-acid to another when one assesses it by means of the formula

$$\Delta\varepsilon_s(T) = \Delta \sum_i Q_i^2 / 2r_i.$$

## C. THE ACID-BASIC CHARACTER OF CONJUGATED HYDROCARBONS IN THE GROUND STATE AND IN THEIR FIRST EXCITED STATES

Not many hydrocarbons possess sufficient acidity for their values of $p_K$ to be directly measurable.

Fluoradene

is one of them. Its $p_K$, measured in a 97% aqueous solution of methyl alcohol, is 13.5

[66] and corresponds to the equilibrium

Consequently, the determination of the values of $p_{Ka}$ of hydrocarbons is most often carried out by comparison by studying the equilibrium with an organometallic compound

$$R'Na + R''H \rightleftharpoons R'H + R''Na .$$

Conant and Wheland [67] and McEwen [68] determined numerous values of $p_{Ka}$ in this way. Table XVI contains a few of these.

TABLE XVI

| | | | |
|---|---|---|---|
| Indene. | . | . | 21 |
| Fluorene | . | . | 25 |
| Toluene. | . | . | 37 |

Conant and Wheland have calculated $\Delta \varepsilon_d$ by means of Hückel's method and have shown that this term varies appreciably like $p_K$. More recently, this study has been extended, the conclusions remaining more or less the same [69].

We should also point out that Streitwieser *et al.* [70] have proposed fresh values for the $p_{Ka}$'s of hydrocarbons.

In a very acid medium the hydrocarbons behave, on the other hand, like bases. Thus in the presence of hydrofluoric acid anthracene is converted partially into the ion:

Making use of reactions of this type, Mackor *et al.* [71] have measured the $p_K$ corresponding to the equilibrium

$$ArH + H^+ \rightleftharpoons ArH_2^+$$

for numerous alternant hydrocarbons.

They have assumed that the value of $p_K$ was principally determined by $\Delta \varepsilon_d$ and that for this reason the proton was preferably fixed on the carbon corresponding to the smallest $\Delta \varepsilon_d$. One will observe that when the proton is fixed on one of the carbons, two electrons belonging initially to the delocalised bond are localised along the new CH bond which is formed.

If in the initial hydrocarbon there was a delocalised bond with $n$ electrons extending over $n$ centres, the ion will contain a delocalised bond with $n-2$ electrons extending over $n-1$ centres. The difference $\Delta\varepsilon_d$ is frequently designated by the name of *electrophilic localisation energy of the carbon attacked.*

Mackor *et al.* have verified that there existed a very satisfactory linear relationship between the $p_K$ of alternant hydrocarbons and the weakest energy of localisation calculated by means of Hückel's method.

This study was then extended to the case of various substituted derivatives of these hydrocarbons [72].

Furthermore, Dallinga *et al.* [73] have calculated the $\Delta\varepsilon_d$ relating to the ions of alternant hydrocrabons using the approximation of Pariser and Parr. They obtained the relationship

$$p_K = -\left(17\Delta\varepsilon_d/\beta\right) - 3.7$$

in which

$$\beta = -2.06\ \text{eV}.$$

More recently Colpa *et al.* [74], [75] have concerned themselves with the values of $p_K$ of the same hydrocarbons in their first excited state. We will use ArH* to denote the state of excitation under consideration. Besides the equilibrium

$$\text{ArH} + \text{H}^+ \rightleftharpoons \text{ArH}_2$$

one may in fact envisage the equilibrium

$$\text{ArH}^* + \text{H}^+ \underset{K*}{\overset{K}{\rightleftharpoons}} \text{ArH}_2^*.$$

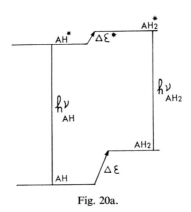

Fig. 20a.

Figure 20a shows clearly that one has

$$\Delta\varepsilon^* = \Delta\varepsilon - \left(h\nu_{\text{AH}} - h\nu_{\text{AH}_2}\right) \tag{1}$$

if $hv_{AH}$ and $hv_{AH_2}$ are used to denote respectively the transition energies of the bands $0 \to 0$ for the hydrocarbon and its ion and if $\Delta\varepsilon^*$ represents the difference between the energy of the excited ion and that of the excited state of the hydrocarbon.

Assuming that the ratio of the partition functions is the same for the fundamental state and for the excited state, one gets [76]

$$M\left(p_{K^*} - p_K\right) = \frac{1}{\chi T}\left(hv_{AH_2} - hv_{AH}\right). \tag{2}$$

Since as a rule $hv_{AH_2}$ is greater than $hv_{AH}$, the value of $p_{K^*}$ for the excited state must be smaller than that of the ground state.

In other words, a *conjugated hydrocarbon must be much more basic in its first excited state than in the ground state.*

If now with Copla *et al.* one assumes that

$$\Delta\varepsilon^* - \Delta\varepsilon = \Delta\varepsilon_d^* - \Delta\varepsilon_d$$

one arrives at the relationship

$$\Delta\varepsilon_d^* = \Delta\varepsilon_d - \left(hv_{AH} - hv_{AH_2}\right). \tag{3}$$

Colpa *et al.* have calculated the values of $\Delta\varepsilon_d^*$ from this relationship by determining $\Delta\varepsilon_d$ and $hv_{AH_2}$ by means of the method of Pariser and Parr, but using the experimental values of $hv_{AH}$. They have on the other hand evaluated the values of $p_{K^*}$ of the excited states from Weller's relationship.

Table XVII makes it possible to compare the results so obtained.

TABLE XVII

| | $-\dfrac{\Delta\varepsilon d^*}{\beta}$ | $p_{K^*}$ |
|---|---|---|
| Anthracene (apex 9) | $-0.15$ | $-10$ |
| Pyrene (apex 3) | $-0.05$ | $-15$ |
| Perylene (apex 1) | $-0.16$ | $-18$ |
| 3,4-benzopyrene (apex 5) | $-0.31$ | $-18$ |

It is difficult to draw definitive conclusions from this study. Many hypotheses have been made which doubtless are not justified. One assumes in particular that in the excited state one will have a true thermodynamic equilibrium which will be established between the hydrocarbon and the ion. It does not seem that one possesses experimental proof on this subject. Moreover, even if it were so, there would still be no reason to suppose that the ion which predominates in the excited state corresponds to a fixation of the proton on the apex preferably attacked in the ground state.

It would be interesting to see whether it is the most negative apex of the excited state of AH which corresponds to the weakest $\Delta\varepsilon_d^*$. Finally one is perhaps wrong in not taking into account $\Delta\varepsilon_s^*(T)$.

Ehrenson [77] recently devoted himself to an examination in depth of the effect of the methyl substituent on the value of $p_k$ of alternant hydrocarbons. His study is based on the experimental data of McCauley *et al.* [78, 79] Kilpatrick and Luborsky [80], Mackor *et al.* [81]. We will point out that the effect of methylation on $p_K$ is very noteworthy; six units of $p_K$ separate the case of benzene from that of hexamethyl-benzene.

Ehrenson first of all calculated $\Delta\varepsilon_d$ by introducing the *hyperconjugation associable with methyl* according to the technique of Muller, Pickett and Mulliken [82]. This method consists in assuming that the delocalised bond extends to the methyl group by means of an orbital $2p_z$ centred on its carbon atom and a *'quasi-$\pi$ orbital'* formed on three atomic orbitals $1_s$ of the hydrogen atoms. 'Quasi-$\pi$' is used to qualify an orbital which is essentially positive on one side of the plane of the molecule and nega-tive on the other side without the negative part of the function being necessarily symmetrical with the positive part as is the case for a true $\pi$ orbital. One also assumes that the integrals $\beta$ are proportional to the overlap integrals and one adopts the relationships

$$\alpha_{H_2} = \alpha^0 - 0.3\beta_0$$

for the quasi-$\pi$ orbital and

$$\alpha_C = \alpha^0 - 0.075\beta_0$$

for the carbon of the methyl.

Ehrenson also took into account a certain *charge effect*, that is to say a relationship between the values of $\alpha_j$ and the electronic charge $q_j$ of the atom $j$ defined by the relationship

$$\alpha_j = \alpha_j^0 + \omega\beta_0(1 - q_j)$$

and in consequence of this he carried out the calculations by iteration.

Table XVIII makes it possible to compare with the experimental constants K related to that of paraxylene the same relative constants calculated whilst only taking into account $\Delta\varepsilon_d$.

In view of the imprecision of the experimental data, one may consider that the hyper-conjugation is sufficient to take into account the essence of the phenomenon.

TABLE XVIII

|  | Calculated values | Measured values |  |
|---|---|---|---|
| Benzene | 0.004 | 0.09 |  |
| Metaxylene | 34 | 20 to | 26 |
| Durene | 98 | 120 to | 140 |
| Prehnitene | 180 | 170 to | 400 |
| Mesitylene | 1 900 | 2 800 to 13 000 |  |
| Hexamethylbenzene | 11 000 | 89 000 to 97 000 |  |

Ehrenson tried to see whether from another point of view it would be possible also to obtain an agreement between the theoretical results and the experimental data by restricting oneself to representing the effect of the methyl by means of a simple inductive effect, that is to say by modifying the integral $\alpha$ of the ring carbon which carries the methyl group.

The calculation provides a positive reply to this question, but on condition that one modifies $\alpha$ by at least 0.5 eV, which Ehrenson finds unreasonable. However, Flurry and Lykos [83] took up this calculation of the inductive effect of the methyl once again within the framework of the method of the auto-coherent field and the approximations of Pariser and Parr and concluded that without introducing exaggerated values for the parameters it is possible to take the experimental values of $p_K$ perfectly into account. They also noted that their process made it possible to calculate correctly the distribution of the spin density in the negative ion of methylbenzene. Having been encouraged in this way the authors calculated $\Delta\varepsilon_d^*$ for the first triplet excited state of the methylbenzenes and published the list of the values of $p_K$ thus envisaged by theory for these excited states.

## D. STRENGTH OF HETERO-ATOMIC CONJUGATED MOLECULES
### IN THE FUNDAMENTAL STATE

Numerous authors have applied the methods of wave mechanics to the study of the values of $p_K$ of the aza derivatives of conjugated hydrocarbons and various corresponding substituted molecules.

The first works [84–88] were limited to the calculation of $\Delta\varepsilon_d$ by means of naive methods sometimes even used in simplified forms. More recently [89–92] $\Delta\varepsilon_d$ was

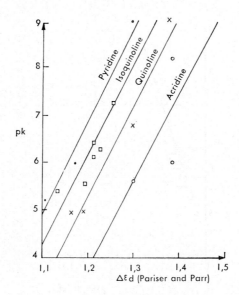

Fig. 20b.   • Derivatives of pyridine;   □ Derivatives of isoquinolein;   × Derivatives of quinolein;
○ Derivatives of acridine.

calculated by using the self-consistent field method in the approximations of Pariser and Parr. In point of fact the qualitative aspect of the results does not depend a great deal on the elaboration of the method. Figure 20b shows the relationship which exists between the values of $p_K$ of the amino derivatives of pyridine, quinolein, isoquinolein, acridine and the values of $\Delta\varepsilon_d$ calculated according to Pariser and Parr's approximation.

There have been omitted from this figure the points corresponding to molecules where the amine group is situated in the ortho or in the peri position of the hetero atom because of the obvious occurrence of an interaction between the hydrogen atoms of the $NH_2$ group and the free pair of the heterocyclic nitrogen corresponding to a term $\Delta\varepsilon_{n\cdot l}$ which is difficult to evaluate.

As will be seen, the points relating to one and the same family of molecules are located close to one and the same straight line. We have seen that the term connected with the phenomena of solvation $\Delta\varepsilon_s(T)$ depends to a large extent on the size of the molecules, and this is why Chalvet et al. (loc. cit.) thought that this segregation into families could be due to the influence of this term. Calculating the latter from Hoijtink's formula

$$\Delta\varepsilon_s(T) = -\Delta\sum_i \frac{Q_i Q_j}{2r_{ij}}\left(1 - \frac{1}{D}\right)$$

they obtain Figure 21 showing that the segregation into families disappears if one represents the variation of $p_K$ as a function of the sum

$$\Delta\varepsilon_d + \Delta\varepsilon_s(T).$$

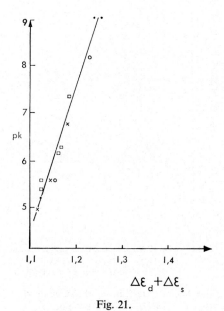

Fig. 21.

Kende studied the value of $p_K$ of carbonyl compounds such as benzaldehyde, eucarvone, diphenylcyclopropenone, tropone, fuchsone, etc. Many of these compounds are derived from non-alternant hydrocarbons.

Kende found that the relationship between $p_K$ and $\Delta\varepsilon_d$ is better when the latter magnitude is calculated by the self-consistent field method in Pariser and Parr's approximation by comparison with the results obtained by means of Pauling and Wheland's method [93].

We would also point out that the value of $p_K$ of different organic molecules possessing an acid character has also formed the subject of analogous studies [94].

## E. THE EFFECT OF ELECTRONIC EXCITATION ON THE ACID/BASIC PROPERTIES OF HETERO-ATOMIC CONJUGATED MOLECULES

Coulson and Jacobs [95] studying the distribution of the electronic charges in the ground state and the lowest excited states of aniline were, it appears, the first to predict that such a molecule would have to be much less basic in its excited states than in the ground state.

At the same time Förster [96] demonstrated that it frequently happens that the *acid/basic equilibrium has time to become established during the life of the excited molecule*, so that the fluorescence spectrum provides a means of determining the corresponding $p_K$. Weller's relationship, of which we have already spoken, provides another means of access to this $p_K$ and frequently the values obtained by these two methods agree fairly well with one another.

Furthermore it has been observed that the speed of transfer of a proton of the solvent to an excited basic receptor does not depend very much on the latter [97], so that the $p_K$ also provides a measure of the speed of the transfer of the proton of the excited molecule to the solvent.

Förster [98], observed in particular that for basic pH's up to the value 2, a solution of 3-aminopyrene only contains in their ground states the neutral molecules, and the $ArNH_3^+$ ions only appear below a pH value of 2.

The behaviour of the excited entities in first singlet state appears to be quite different. For values of pH higher than 12 it is the $ArNH^-$ ions which predominate as if in this excited state $ArNH_2$ behaved as an acid and reacted according to the reaction

$$ArNH_2 \rightleftharpoons ArNH^- + H^+.$$

This acid character of the excited state of $ArNH_2$ is reinforced by the absence of the fluorescence spectrum of the $ArNH_3^+$ ions, even at pH zero.

In order to interpret this phenomenon, Sandorfy [99] calculated the distribution of the electronic charges in aniline (ground state and first excited state), taking into account both the delocalised system and the localised bonds. He observed that, whilst the total charge of the nitrogen is negative in the ground state, it becomes positive in the first excited state. One must therefore expect that it will lose its basic properties. It seems very probable that analogous calculations carried out in regard to 3-aminopyrene would give qualitatively comparable results.

Jaffé *et al.* [100] also observed the existence of a very satisfactory relationship between the difference of $p_K$ which characterises two electronic states of azobenzene or azoxybenzene and the difference of the charges carried either by the nitrogen or by the oxygen in these states.

Jackson and Porter [101] determined the $p_K$ values of conjugated molecules in their first triplet state (Table XIX).

TABLE XIX

|  | $p_K$ (ground state) | $p_K$ (first excited singlet state) | $p_K$ (first triplet state) |
|---|---|---|---|
| β-naphthol | 9.46 | 2.8 | 8.1 |
| β-naphthylamine | 4.1 | −2 | 3.3 |
| Acridine | 5.5 | 10.6 | 5.6 |

It will be seen that the $p_K$ of the triplet state is much closer to that of the ground state than that of the excited singlet state. Murrell [102] gave an interpretation of this fact by stating that if one is dealing with a molecule of the type $ArNH_2$, for example, considering the union by perturbation of Ar and $NH_2$, the orbitals corresponding to a charge transfer are further away energetically from the triplet state than from the first excited singlet state.

Linnett [103] also gave a qualitative interpretation of this phenomenon.

To emphasise the variety of phenomena which can be observed, we will finally point out that Haylock *et al.* [104] observed that in the case of 3-hydroxyquinolein the OH group becomes more acid and the nitrogen more basic in the excited state as compared with what happens in the ground state. Pauling and Wheland's method made it possible for them to interpret this phenomenon.

## 7. The Association Constants

We will now study a certain number of equilibrium constants corresponding to the association of two molecular entities:

$$A + B \rightleftharpoons A, B.$$

There exist numerous classes of such equilibria such as, for example, those which correspond to the formation of complexes by charge transfer, the association of free radicals, the association between an acceptor and a proton donor.

A. ASSOCIATION BETWEEN AN ACCEPTOR AND A PROTON DONOR

We must emphasise a little this last-mentioned type of equilibrium because it is not without its connections with the acid-basic equilibria which we have just studied and the comparison between the two types of equilibrium is very instructive.

The idea of association is very old because, as far back as 1884, Raoult [105] used this idea to interpret the results of his cryoscopic studies. Later on Dolezalek [106] explained quantitatively by the formation of a molecular chloroform/acetone complex the discrepancies from the law of Raoult presented by the partial pressures of mixtures of these molecules. Hildebrand *et al.* [107], however, observed that nonspecific Van der Waals' forces would be sufficient in certain cases to explain such phenomena. It should, however, be noted that, according to Bratoz and Martin [108], the Van der Waals' complexes which are formed in certain gaseous mixtures at high pressure a certain specificity.

Latimer and Rodebush [109] explained numerous molecular associations with the help of the notion of a hydrogen bond. Freymann [110] was one of the first to explain with the help of association the variations with the concentration of the frequency of vibration of valency of the group OH of alcohol dissolved in benzene. Errera and Mollet [111] studied in detail other examples and then a large number of works were carried out along this route, particularly in the groups of Professors Josien, Lascombe [112], Champetier [113], Huyskens [114, 115] and Martin [116].

In parallel, use was made of the perturbations produced on the electronic spectra by the hydrogen bond for measuring the association constants between donor and proton acceptor [117–121]. The literature is therefore very rich [122] in experimental data on this subject and we shall have to restrict ourselves only to discussing a few examples; those which enable us to make fruitful comparisons with the values of $p_K$.

Let B be the proton acceptor and HR the proton donor. We will have to compare the equilibria

$$B + H^+ \underset{K}{\rightleftharpoons} BH^+$$

and

$$B + HR \underset{K'}{\rightleftharpoons} B, HR.$$

To fix one's ideas, we will assume that the association is produced thanks to the appearance of a hydrogen bond.

We are led to write the usual expressions of equilibrium constants

$$K = \frac{f_{BH^+}}{f_B f_{H^+}} \exp\left[-\frac{\Delta\varepsilon_v + \Delta\varepsilon_l + \Delta\varepsilon_d + \Delta\varepsilon_{n\cdot l} + \Delta\varepsilon_s(T)}{\chi T}\right]$$

$$K' = \frac{f_{B,HR}}{f_B f_{HR}} \exp\left[-\frac{\Delta\varepsilon_v' + \Delta\varepsilon_l' + \Delta\varepsilon_d' + \Delta\varepsilon_{n\cdot l}' + \Delta\varepsilon_s'(T)}{\chi T}\right]$$

and to discuss the respective importance of the various terms. Let us choose the example of conjugated molecules. We have seen that then the values of $p_K$ depend mainly on

$$\Delta\varepsilon_d, \ \Delta\varepsilon_s(T) \ \text{ and } \ \Delta\varepsilon_{n\cdot l}.$$

(1) We must expect that the $\Delta\Delta\varepsilon_d'$ will be very much lower than the $\Delta\Delta\varepsilon_d$ because the distance which separates the hydrogen of the acceptor in B, HR is obviously dis-

tinctly greater than in $BH^+$ and moreover the positive charge of this hydrogen atom runs the risk of being lower in B, HR than in $BH^+$.

*One must therefore expect that the association constants will be less sensitive than the values of $p_K$ to the effects of structure which can be represented by the $\Delta\varepsilon_d$'s.* This attenuation, however, may be partly compensated for by the existence in the equation of $K'$ of a $\Delta\varepsilon_1'$ representing the energy of formation of the hydrogen bond which may vary like the $\Delta\varepsilon_d$'s. In point of fact the stability of the hydrogen bond may vary like the donor power of the acid and like the acceptor power of the base.

(2) Since the association constants are measured in solvents which are not very polar, one will generally have:

$$\Delta\Delta\varepsilon_s'(T) \ll \Delta\Delta\varepsilon_s(T).$$

(3) The values of $\Delta\Delta\varepsilon'_{n \cdot l}$ may on the other hand be just as important as those of $\Delta\Delta\varepsilon_{n \cdot l}$, but sometimes they may behave differently, as effects of direct steric hindrance may be produced between B and R, HR.

All in all, one must predict that the existence of a relationship between the values of $p_K$ and the association constants will be very precarious particularly when the values of $p_K$ depend to a large extent on solvation.

Table XX summarises the results of Bonnet (*loc. cit.*) who, by studying the ultraviolet spectra, measured the association constants of phenols and aliphatic amines dissolved in normal heptane.

TABLE XX

Values of the constants $K'$ at 25 °C

|  |  |  | Paracresol | Phenol | α-naphthol |
|---|---|---|---|---|---|
|  | $10^6 K$ | $10^{10} K$ | 0.65 | 1.08 | 1.40 |
| Triethylamine | 435 |  | 55 | 86 | 117 |
| Tributylamine | 1282 |  | 83 | 128 | 162 |

As expected, the association constants vary less distinctly than the values of $p_K$, but the variations here remain quite parallel. The author interprets this observation by the consideration of the term $\Delta\varepsilon_l'$.

Table XXI contains the results of Sprecher (*loc. cit.*). This worker, according to the method of Fuson, Pineau and Josien, studied the association of methyl alcohol with different mono-aza-aromatic bases in carbon tetrachloride.

Once again it would appear that $K'$ varies less than $K$. The influence of the steric encumbrance is obvious for $K'$ as in the case of $K$ in the case of 4-azaphenanthrene. But this time there is practically no relationship between the values of $p_K$ and the association constants. We are not too surprised by this because it is precisely a case where we know that the effects of solvation play a great part in the determination of the $p_K$'s.

The members of Professor Huysken's group have studied the effect of substitution

TABLE XXI

|  | $p_K$ | $\log K'$ |
|---|---|---|
| 4-azaphenanthrene | 4.25 | 0.204 |
| Quinolein | 4.94 | 0.544 |
| Isoquinolein | 5.14 | 0.580 |
| 1-azaphenanthrene | 5.15 | 0.672 |
| Pyridine | 5.23 | 0.491 |
| Acridine | 5.60 | 0.690 |

on the association constants and on the values of $p_K$. In greater detail, they measured the constants $\varrho$ of Hammett

$$\varrho = \frac{1}{\sigma} \frac{LK'_{subst}}{LK'} \qquad \varrho_i = \frac{1}{\sigma} \frac{LK_{subst}}{LK}$$

the index subst. referring to the constants of substituted molecules and $\sigma$ being the coefficient of Hammett relating to the substituent.

We therefore have

$$\frac{\varrho}{\varrho_i} = \frac{\Delta Lf - (1/\chi T)\left[\Delta\Delta\varepsilon'_v + \Delta\Delta\varepsilon'_l + \Delta\Delta\varepsilon'_d + \Delta\Delta\varepsilon'_{n\cdot l} + \Delta\Delta\varepsilon'_s(T)\right]}{\Delta Lf' - (1/\chi T)\left[\Delta\Delta\varepsilon_v + \Delta\Delta\varepsilon_l + \Delta\Delta\varepsilon_d + \Delta\Delta\varepsilon_{n\cdot l} + \Delta\Delta\varepsilon_s(T)\right]}.$$

Now Huyskens et al. obtained*:

$\varrho/\varrho_i = 0.3$ for the substituted phenol/aniline system;

$= 0.4$ for the substituted phenol/pyridine system;

$= 0.76$ for the substituted phenol/triethylamine system;

$= 0.24$ for the phenol/substituted aniline system;

$= 0.21$ for the phenol/substituted pyridine system;

As expected, $\varrho/\varrho_i$ is always lower than unity: $K'$ is less sensitive than $K$ to the effects of substituents. Moreover, as Professor Huyskens himself wrote to me, "this ratio is greater, the stronger the hydrogen bond becomes, that is to say the more the proton approaches the acceptor."

B. OTHER EXAMPLES

We have already given a few references on the complexes formed by charge transfer. We will content ourselves with referring the reader to the book *Quantum Chemistry* (Daudel, Lefebvre, and Moser, Interscience Publ., 1949, p. 279) on the subject of the study of the stability of free radicals.

## 8. Tautomeric Equilibria: Application to the Study of the Mechanism of Mutations

Numerous tautomeric equilibria have been studied by means of the methods of

* Personal communication.

quantum chemistry (see for example, Daudel, Lefebvre, and Moser, *Quantum Chemistry*, Interscience Publ., 1959, p. 277).

To conclude this chapter we will examine an example of biochemical interest. The formula below represents the classic coupling of cytosine and guanine which is effected in nucleic acids

Cytosine is present there in the amine form.

It is known that this base also exists in the imine form

so that one is brought to consider the equilibrium between tautomers

A. Pullman [123] found that for this equilibrium $\Delta\varepsilon_d$ is $-0.13\beta$. In other words, the energy of the delocalised bond is greater in the amine form than in the imine form. This would be one of the causes of the rarity of the imine form and consequently of the abundance of the amine form in nucleic acids.

The imine form of cytosine may give rise to a pairing with adenine

Such a pairing which is quite abnormal may, if it occurs, involve an error of duplication of desoxyribonucleic acid, and therefore a *mutation*.

Now A. Pullman points out that in the first excited electronic state $\Delta\varepsilon_d$ is reduced to $-0.01\beta$. In such a state the imine form runs the risk therefore of becoming more abundant, which may be a possible explanation of the mutagenic effect of ultraviolet light.

## References

[1] After: E. Mulder and I. Muller Rodloff, *Ann.* **517** (1935) 134; E. Muller and E. Hertel, *Ann.* **554** (1943) 213; **555** (1944) 157;
E. Muller and H. Pfauz, *Ber.* **74** (1941) 1051–1075.
[2] See on this subject: D. D. Eley and M. G. Evans, *Trans. Far. Soc.* **34** (1938) 1112.
[3] M. Born, *Z. Phys.* **1** (1920) 45.
[4] W. M. Latimer, K. S. Pitzer, and C. M. Slansky, *J. Chem. Phys.* **7** (1939) 108.
[5] B. Noyes, *J. Am. Chem. Soc.* **86** (1962) 513; **86** (1964) 971.
[6] E. Glueckhauf, *Trans. Far. Soc.* **60** (1964) 572.
[7] R. Stokes, *J. Am. Chem. Soc.* **86** (1964) 979.
[8] G. J. Hoijtink, E. de Boer, P. H. Van der Meij, and W. P. Weijland, *Rec. Trav. Chim. Pays-Bas* **75** (1956) 487.
[9] D. D. Eley and M. G. Evans, *Trans. Far. Soc.* **34** (1938) 1112.
[10] H. A. Laitinen anb S. Wawzonek, *J. Chem. Soc.* **64** (1942) 1765.
[11] S. Wawzonek and H. A. Laitinen, *J. Chem. Soc.* **64** (1942) 2365.
[12] I. Bergman, *Trans. Far. Soc.* **50** (1954) 289.
[13] G. J. Hoijtink, J. Van Schooten, E. de Boer, and W. I. Aalbersberg, *Rec. Trav. Chim. Pays-Bas* **73** (1954) 355.
[14] Delabray, *New Experimental Methods in Electrochemistry*, Interscience, New York, 1954.
On the theory of polarography one can also read:
J. E. Page, *Quart. Rev.* **6** (1952) 262;
I. M. Kolthoff and J. J. Lingane, *Polarography*, Interscience, New York, 1952;
L. Meites, *Polarographic Techniques*, Interscience, New York, 1955;
G. W. C. Milner, *The Principle and Application of Polarography and other Electroanalytical Processes*, Longmans Green, New York, 1957;
H. W. Nürnberg, *Angew. Chem.* **72** (1960) 433.
A table of semiwave potentials is to be found in:
K. Schwabe, *Polarographie und chemische Konstitution organischer Verbindungen*, Ac. Verlag, Berlin, 1957.
[15] A. Maccoll, *Nature* **163** (1949) 178.
[16] A. Pullman, B. Pullman, and C. Berthier, *Bull. Soc. Chim. Fr.* **17** (1950) 591.
[17] G. J. Hoijtink and J. Van Schooten, *Rec. Trav. Chim. Pays-Bas* **71** (1952) 1089; **72** (1953) 691, 903.
[18] G. J. Hoijtink, *Rec. Trav. Chim. Pays-Bas* **74** (1955) 1525.
[19] F. A. Matsen, *J. Chem. Phys.* **24** (1956) 602;
[20] R. M. Hedges and F. A. Matsen, *J. Chem. Phys.* **28** (1958) 950.
[21] I. Jano, *Thèses Sciences*, Paris 1965.
[22] J. Hoyland and L. Goodman, *J. Chem. Phys.* **36** (1962) 12, 21.
[23] W. E. Wentworth and R. S. Becker, *J.A.C.S.* **84** (1962) 4263.
[24] R. S. Becker and W. E. Wentworth, *J.A.C.S.* **85** (1963) 2210.
[25] A. Watson and F. Matsen, *J. Chem. Phys.* **18** (1950) 1305.
[26] I. Bergman, *Trans. Far. Soc.* **50** (1954) 829; **52** (1956) 690.
[27] S. Basu and R. Bhattacharya, *J. Chem. Phys.* **25** (1956) 596.
[28] J. I. Fernandez-Alonso and R. Domingo, *Nature* **179** (1957) 829.
[29] P. H. Given, *Nature* **181** (1958) 1001.
[30] H. Lund, *Acta. Chem. Scand.* **11** (1957) 1323.
[31] G. J. Hoijtink, *Rec. Trav. Chim. Pays-Bas* **77** (1958) 555.

[32] W. I. J. Aalbersberg and E. L. Mackor, *Trans. Far. Soc.* **56** (1960) 1351.
[33] N. S. Hush and J. A. Pople, *Trans. Far. Soc.* **51** (1955) 600.
[34] A. Streitwieser, *Molecular Orbital Theory for Organic Chemists*, Wiley, 1961, p. 183.
[35] G. Anthoine, G. Coppens, J. Nasielski, and E. Van der Donckt, *Bull. Soc. Chim. Belg.* **73** (1964) 65.
[36] S. Basu and J. N. Chandhuri, *Nature* **180** (1957) 1473.
[37] R. Zahradnik and K. Bocek, *Coll. Czech. Chem. Comm.* **26** (1961) 1733.
[38] R. W. Schmid and E. Heilbronner, *Helv. Chim. Acta* **37** (1954) 1453.
[39] G. Klopman and J. Nasielski, *Bull. Soc. Chim. Belge* **70** (1961) 490.
[40] G. Giacometti, *La ricerca scientifica* **27** (1957) 1146.
[41] G. Giacometti, *La ricerca scientifica* **27** (1957) 1489.
[42] G. J. Hoijtink, E. de Boer, P. H. Van der Meij, and W. P. Weijland, *Rec. Trav. Chim. Pays-Bas* **75** (1956) 487.
[43] T. Fueno, T. Ree and H. Eyring, *J. Am. Chem. Soc.* **63** (1959) 1940.
[44] G. E. K. Branch and M. Calvin, *The Theory of Organic Chemistry*, Prentice Hall, 1941, p. 305.
[45] M. E. Diatkina and J. Syrkin, *Acta Physiocochim. URSS* **21** (1946) 921.
[46] M. G. Evans, *Trans. Far. Soc.* **42** (1946) 113.
[47] E. Berliner, *J. Am. Chem. Soc.* **68** (1946) 49.
[48] C. J. P. Sprint, *Chem. Week* **43** (1947) 544.
[49] G. B. Bonino and M. Rolla, *Atti. Accad. naz. Lincei, Mem. Classe Sci. Fis. Mat. e nat.*, **4** (1948) 25, 273.
[50] P. G. Carter, *Trans. Far. Soc.* **45** (1949) 597.
[51] M. G. Evans, J. Gergely, and J. de Heer, *Trans. Far. Soc.* **45** (1949) 312.
[52] M. G. Evans and J. de Heer, *Trans. Far. Soc.* **47** (1951) 801; *Quart. Revs.* **4** (1950) 94.
[53] V. Gold, *Trans. Far. Soc.* **46** (1950) 109.
[54] J. Deschamps, *Thèse Sciences*, Bordeaux 1956.
[55] R. Le Bihan, *Thèse 3° cycle Sciences*, Paris 1965.
[56] A. Pullman, *Cont. Rend. Acad. Sci.* **253** (1961) 1210.
[57] A. Pullman, *Tetrahedron* **19** (1963) 441.
[58] M. E. Pullman and C. P. Colowick, *Fed. Proc.* **12** (1953) 255.
[59] M. E. Pullman, A. San Pietro, and C. P. Colowick, *J. Biol. Chem.* **206** (1954), 129.
[60] Vennesland, *The Physical Chemistry of Enzymes*, 1955, p. 240.
[61] B. M. Anderson and N. O. Kaplan, *J. Biol. Chem.* **234** (1959) 1226.
[62] N. O. Kaplan and M. M. Ciotti, *J. Biol. Chem.* **221** (1956) 823.
[63] G. Del Ré, *J. Chem. Soc.* (1958) 4031; *Electronic Aspects of Biochemistry*, Ac. Press, New York, 1964, p. 221.
[64] G. Del Ré, B. Pullman, and T. Yonezawa, *Biochim. Biophys. Acta* **75** (1963) 153.
[65] T. Yonezawa, G. Del Ré, and B. Pullman, *Bull. Chem. Soc. Japan* **37** (1964) 985.
[66] H. Rapoport and G. S. Smolinsky, *J. Am. Chem. Soc.* **82** (1960) 934.
[67] J. B. Conant and G. W. Wheland, *J. Am. Chem. Soc.* **54** (1932) 1212.
[68] W. K. McEwen, *J. Am. Chem. Soc.* **58**, (1936) 1124.
[69] A. Streitwieser, *Tetrahedron Letters*, No. 6 (1960) 23.
[70] A. Streitwieser, W. C. Langworthy, and J. I. Brauman, *J. Am. Chem. Soc.* **85** (1963) 1761.
[71] E. L. Mackor, A. Hofstra, and J. H. Van der Waals, *Trans. Far. Soc.* **54** (1958) 66.
[72] J. P. Colpa, C. Maclean, and E. L. Mackor, *Tetrahadron* **19** (1963) 65.
[73] G. Dallinga, A. A. Verrijn Stuart, P. J. Smit, and E. L. Mackor, *Electrochem.* **61** (1957) 1019
[74] E. L. Mackor, G. Dallinga, J. H. Kruizinga, and A. Hofstra, *Rec. Trav. Chim. Pays-Bas* **75** (1956) 836.
[75] E. L Mackor, A. Hofstra, and J. H. Van der Waals, *Trans. Far. Soc.* **54** (1958) 186.
[76] A. Weller, *Z. Electrochem.* **61** (1957) 957; *Disc. Far. Soc.* **27** (1959) 28.
[77] S. Ehrenson, *J. Am. Chem. Soc.* **83** (1961) 4493; **84** (1962) 2681.
[78] D. A. McCaulay, B. H. Shoemaker, and A. P. Lien, *Ind. Eng. Chem.* **42** (1950) 2103.
[79] D. A. McCaulay and A. P. Lien, *J. Am. Chem. Soc.* **73** (1951) 2013.
[80] M. Kilpatrick and F. E. Luborsky, *J. Am. Chem. Soc.* **75** (1953) 577.
[81] E. L. Mackor, A. Hofstra, and J. H. Van der Waals, *Trans. Far. Soc.* **54** (1958) 186.
[82] N. Muller, L. W. Pickett and R. S. Mulliken, *J. Am. Chem. Soc.* **76** (1954) 4770.
[83] R. L. Flurry and P. G. Lykos, *J. Am. Chem. Soc.* **85** (1963) 1033.

[84]  J. J. Elliott and S. F. Mason, *J. Chem. Soc.* (1959) 2352.

[85]  J. Ploquin, *Compt. Rend. Acad. Sci.* **226** (1948) 2140;
      R. Daudel, *Compt. Rend. Acad. Sci.* **227** (1948) 1241.

[86]  R. Daudel and O. Chalvet, *J. Chim. Phys.* **46** (1949) 332.

[87]  H. C. Longuet-Higgins, *J. Chem. Phys.* **18** (1950) 275.

[88]  O. Chalvet, M. Pages, M. Roux, N. P. Buu-Hoi, and R. Royer, *J. Chim. Phys.* **51** (1954) 548.

[89]  A. Pullman and T. Nakajima, *J. Chim. Phys.* **55** (1958) 793.

[90]  O. Chalvet, R. Daudel, and R. Peradejordi, *J. Chim. Phys.* **59** (1962) 709.

[91]  F. Peradejordi, *Cahiers de Phys.* **17** (1963) 393.

[92]  O. Chalvet, M. J. Huron, and F. Peradejordi, *Compt. Rend. Acad. Sci.* **259** (1964) 1631.

[93]  A. Kende, *In the Application of Wave Mechanical Methods to the Study of Molecular Properties*, (R. Daudel, ed.) *Adv. in Chemical Physics* **8** (1965) 133.

[94]  See, for example: A. and B. Pullman, *Les Théories Electroniques de la Chimie Organique* Masson, Paris, 1952.

[95]  C. A. Coulson and J. Jacobs, *J. Chem. Soc.* (1949) 1893.

[96]  T. Förster, *Naturwiss.* **36** (1949) 186.

[97]  M. Eigen, W. Kruse, G. Maass, and L. DeMaeyer, *Progress in Reaction Kinetics*, Vol. 2, Pergamon Press, Oxford, 1964, 287.

[98]  T. Förster, *Z. Elektrochem.* **54** (1950) 42, 531.

[99]  C. Sandorfy, *Comt. Rend. Acad. Sci.* **232** (1951) 617.

[100]  H. H. Jaffé, D. L. Beveridge, and H. L. Jones, *J. Am. Chem. Soc.* **86** (1964) 2932.

[101]  G. Jackson and G. Porter, *Proc. Roy. Soc.* **260** (1961) 13.

[102]  J. N. Murrell, *The Theory of Electronic Spectra of Organic Molecules*, Methuen, London, 1964.

[103]  J. W. Linnett, *Electronic Structure of Molecules*, Methuen, London, 1964.

[104]  J. C. Haylock, S. F. Mason, and B. E. Smith, *J. Chem. Soc.* (1963) 4897.

[105]  F. M. Raoult, *Ann. Chim. Phys.* **2** (1884) 66.

[106]  F. Dolezalek, *Z. Physik. Chem.* **64** (1908) 727.

[107]  J. H. Hildebrand and R. L. Scott, *The Solubility of Non Electrolytes* (3rd ed.), Reinhold Publ., New York, 1950.

[108]  S. Bratoz and M. L. Martin, *J. Chem. Phys.* **42** (1965) 1051.

[109]  W. M. Latimer and W. H. Rodebush, *J. Am. Chem. Soc.* **42** (1920) 1419.

[110]  R. Freymann, *Ann. Phys.* **20** (1933) 243.

[111]  J. Errera and P. Mollet, *Compt. Rend. Acad. Sci.* **204** (1937) 259.

[112]  See for example: N. Fuson, P. Pineau, and M. L. Josien, *J. Chim. Phys.* (1958) 454; P. Pineau, *Thèse*, Bordeaux 1961; F. Cruège, *Thèse*, Bordeaux 1963.

[113]  See for example: C. Quivoron, *Thèse*, Paris, 1965.

[114]  See for example: A. M. Dierckx, P. Huyskens, and Th. Zeegers-Huyskens, *J. Chim. Phys.* (1965), 336;

[115]  P. Huyskens, *Industr. Chim. Belge* **30** (1965), 801.

[116]  N. Sprecher, Dissertation, Brussels 1961.

[117]  N. D. Coggeshall and G. M. Lang, *J. Am. Chem. Soc.* **70** (1948) 3283.

[118]  S. Nagakura and H. Baba, *J. Am. Chem. Soc.* **74** (1962) 5693.

[119]  S. Nagakura and M. Goutermann, *J. Chem. Phys.* **26** (1957) 881.

[120]  H. Baba, *Bull. Res. Inst. Appl. Elect.* **9** (1957), Nos. 2 and 3.

[121]  M. Bonnet, *Thèse*, Marseilles, 1960.

[122]  See also for example the numerous works of Gramstad *et al.* in *Acta Chem. Scand.* 1960, 1961, 1962, 1963 and in *Spectrochimica Acta* (1964, 1965), and those of Joesten and Drago, *J. Am. Chem. Soc.* (1962).

[123]  A. Pullman, *Electronic Aspects of Biochemistry*, Ac. Press, New York, 1964, p. 135.

# CHAPTER IV

# RATE CONSTANTS OF A FEW REACTIONS OF
# ORGANIC CHEMISTRY

## 1. Introduction

After analysing in detail the nature of an equilibrium constant and the role of the different factors which determine it we are now ready to examine the methods which make it possible to estimate theoretically the speed constants of chemical reactions as complex as those of organic chemistry.

We will first of all concern ourselves with reactions which are supposed to be adiabatic and take place from molecules situated in their fundamental electronic state. We will also assume that the theory of the transition state is valid. The comparison of the results so obtained and the data from experience will, it appears to us, constitute the best way of judging the value of the postulation.

Let us therefore consider an elementary process of a reaction which in the gaseous phase would boil down to a bimolecular impact such as

$$A + B \overset{k}{\to} C + D + \cdots.$$

According to the theory of the state of transition we will make the following equation correspond to it

$$A + B \overset{K^{\neq}}{\to} M^{\neq} \overset{k^{\neq}}{\to} C + D + \cdots$$

in which

$$k = k^{\neq} K^{\neq}$$

(see Section 3 of Chapter I)

If the reaction is effected in solution the impact will not be truly bimolecular, each molecule A or B being at the moment of impact surrounded by a swarm of molecules of solvents. We will take into account the effects of the solvent in the expression of $k^{\neq}$ and in that of $K^{\neq}$ by making use of the method followed in Chapter III because this gave us very suitable results. We will therefore write $K^{\neq}$ in the form

$$K^{\neq} = \frac{f_{M^{\neq}}^{s}}{f_A^s f_B^s} \exp\left[ - \frac{\Delta\varepsilon_v^{\neq} + \Delta\varepsilon_l^{\neq} + \Delta\varepsilon_d^{\neq} + \Delta\varepsilon_{n\cdot l}^{\neq} + \Delta\varepsilon_s^{\neq}(T)}{\chi T} \right]$$

(after the form given to $K$ in Section 1 of Chapter II).

The speed constant $k^{\neq}$ will itself be written in the form

$$k^{\neq} = (1 + t^s)\, \eta^s \frac{\chi T}{h}$$

(according to the expression given to this magnitude in Section 3 of Chapter I).

The final expression which we will adopt for $k$ will therefore be

$$k = (1 + t^s) \eta^s \frac{\chi T}{h} \frac{f_{M^{\neq}}^s}{f_A^s f_B^s} \exp\left[ - \frac{\Delta\varepsilon_v^{\neq} + \Delta\varepsilon_l^{\neq} + \Delta\varepsilon_d^{\neq} + \Delta\varepsilon_{n \cdot l}^{\neq} + \Delta\varepsilon_s^{\neq}(T)}{\chi T} \right]. \quad (1)$$

This formula shows the eight principal factors of chemical reactivity:

(a) the tunnel effect, taking into account the solvent represented by $1 + t^s$;

(b) the transmission coefficient $\eta^s$ which also depends on the solvent;

(c) the ratio of the partition functions also closely bound up with the solvent;

(d) the change in the energy of vibration of the fundamental states when one passes from the intermediate complex in its state of transition to the initial molecules, namely $\Delta\varepsilon_v^{\neq}$;

(e) the change of energy of the localised bonds $\Delta\varepsilon_l^{\neq}$,

(f) that corresponding to the delocalised system $\Delta\varepsilon_d^{\neq}$,

(g) the change of energy associated with interactions between unbonded atoms $\Delta\varepsilon_{n \cdot l}^{\neq}$,

(h) the change in the energy of solvation $\Delta\varepsilon_s^{\neq}(T)$ in the case of reactions in solution.

## 2. A Few Examples of Studies of Rate Constants in the Field of Saturated Hydrocarbons and Their Derivatives

### A. NUCLEOPHILIC SUBSTITUTIONS $S_{N^2}$

It will clearly be seen that the calculation of a rate constant constitutes a much trickier operation than that of an equilibrium constant. It is first of all necessary to know the mechanism of the reaction so as to be able to fractionate it into its successive or simultaneous elementary processes. In addition it is as well to possess fairly precise indications regarding the structure of the transition states of the different intermediate complexes which mark out these different processes.

The nucleophilic substitutions which are produced on the derivatives of saturated hydrocarbons provide a privileged field of study from this point of view.

Many of these reactions are of the order two, that is to say their speed is simply proportional to the concentration of the reagent and to that of the compound attacked. In this case one may frequently assume that the reaction boils down to a simple bimolecular process. One has shown [1], for example, that the rate of hydrolysis in water (or in an aqueous solution of alcohol) of methyl halides can be written

$$v = k [\text{R}-\text{X}] [\text{OH}^-].$$

It is consequently reasonable to assume for the reaction a process which can be represented by the scheme

$$\text{HO}^- + \text{R}-\text{X} \rightarrow [\text{HO}^{-\delta} \ldots \text{R} \ldots \text{X}^{-\delta'}] \rightarrow \text{HOR} + \text{X}^-.$$

The study of isotopic exchange reactions following exactly the same kinetic have provided first rate information regarding this type of mechanism. Hughes *et al.* [2]

studied from this point of view the exchange of halogen between an alkyl halide and a mineral halide in organic solution. They followed the speed of the exchange using mineral halides containing a radioactive halogen. They also chose alkyl halides containing an asymmetrical carbon in the $C-X$ bond. They were able in this way also to follow the evolution of the natural rotary power during the reaction. One of the reactions thus studied may be formulated as follows

$$Br^{*-} + \begin{matrix} C_6H_{13} \\ \diagdown \\ C-Br \\ \diagup | \\ H \ CH_3 \end{matrix} \xrightarrow[\text{acetone}]{\text{in}} \begin{matrix} C_6H_{13} \\ \diagdown \\ Br^*-C \\ \diagup | \\ H \ CH_3 \end{matrix} + Br^- .$$

They observed that each exchange involves the inversion of the power of rotation of the perturbed molecule.

By generalisation one may therefore assume that such a Walden inversion takes place in the majority of bimolecular nucleophilic substitutions ($S_{N2}$) even when the absence of asymmetry in the carbon of the substitution makes it impossible to check its existence and this hypothesis suggests that for all these reactions one should adopt the following mechanism

$$Y^- + \begin{matrix} R_1 \\ \diagdown \\ R_2-C-X \\ \diagup \\ R_3 \end{matrix} \rightarrow$$

$$Y^{-\delta}\ldots \underset{\overset{\diagup\diagdown}{R_2 \ R_3}}{C} \ldots X^{-\delta'} \rightarrow \underset{\overset{|}{R_3}}{\overset{\overset{R_1}{|}}{Y-C-R_2}} + X^- .$$

We shall also see that theoretical arguments suggest that in the state of transition the three $C-R_i$ bonds will be coplanar, the axis YCX being perpendicular to this plane.

Fukui et al. [3] have proposed a way of simplification for estimating the term $\Delta\varepsilon_l^{\neq}$ of the potential barrier which intervenes in such a reaction.

The basis of their process is the method proposed in 1954 by Sandorfy and Daudel [4] (see second book, p. 130). This method consists in developing the molecular orbitals in the form of linear combinations of atomic orbitals. Since it is a question of representing the saturated bonds, these atomic orbitals are frequently chosen in the form of hybrids. That is why Fukui proposed designating our method by means of the abbreviation LCVO [5].

Fukui et al. also assume that in the initial compound the presence of the carbon attached to the halogen atom may be represented by the introduction of four tetra-hedral hybrids $(sp^3)$ into the orbitals of the bonds starting from this atom.

In the transition state Fukui $et$ $al.$ take into account the presence of this same atom by introducing three trigonal hybrids $sp^2$ (one for each bond $CR_i$) and an orbital $p$ which can be associated both with the bond $Y^{-\delta}...C$ and with the bond $C...X^{-\delta}$.

The authors also imagine that in the bond $C...X^{-\delta}$ the electrons are practically carried by the atom X and that the energy provided by the bond $Y^{-\delta}...C$ is constant for reactions of the same type which cause the intervention of the same reagent Y.

Under these conditions the energy of the transition state may be written as*

$$E\begin{bmatrix} R_1 \\ | \\ C \\ \diagup\diagdown \\ R_2\ R_3 \end{bmatrix} + E[X^-] + \text{const}$$

and the term $\Delta\varepsilon_l^{\neq}$ is written

$$\Delta\varepsilon_l^{\neq} = E\begin{bmatrix} R_1 \\ | \\ C \\ \diagup\diagdown \\ R_2\ R_3 \end{bmatrix} + E[X^-] + \text{const} - E\begin{bmatrix} R_1\diagdown \\ R_2\!-\!C\!-\!X \\ R_3\diagup \end{bmatrix}.$$

The parameters are chosen empirically. The values effectively used by the authors are as follows

$$\alpha_H = \alpha - 0.2\beta \qquad \beta_{CC} = 0.84\beta$$
$$\alpha_F = \alpha + 0.9\beta \qquad \beta_{CH} = 1.1\beta$$
$$\alpha_{Cl} = \alpha + 0.5\beta \qquad \beta_{CF} = 0.5\beta$$
$$\alpha_{Br} = \alpha + 0.45\beta \qquad \beta_{CCl} = 0.65\beta$$
$$\alpha_I = \alpha + 0.40\beta \qquad \beta_{CBr} = 0.58\beta$$
$$\alpha_C = \alpha \qquad\qquad \beta_{CI} = 0.53\beta$$
$$\text{and} \quad \beta' = 0.34\beta$$

(we will recall that $\beta'$ represents the resonance integral between two hybrids coming from the same carbon). The overlap integrals are ignored in accordance with the spirit of Hückel's method.

Tables XXII and XXIII show that the relative speeds of reaction decrease as $\Delta\varepsilon_l^{\neq}$ increases. This term $\Delta\varepsilon_l^{\neq}$ therefore seems to be an element which plays an important part in the determination of the speed of these reactions.

The part played by $\Delta\varepsilon_{n\cdot l}^{\neq}$ is also the subject of an elegant analysis [8]. When one constructs molecular models of the transition states for the reactions studied, one sees in fact that in certain cases considerable steric hindrance can occur. Ingold

* $E[\ ]$ signifies energy of.

TABLE XXII

$$RBr + Cl^- \rightarrow RCl + Br^-$$

| R | $\Delta\varepsilon^{\neq}{}_\iota$ (in $-\beta$ unit) | Experimental relative speed [6] |
|---|---|---|
| Methyl | 0.74 | 100 |
| Ethyl | 0.815 | 1.97 |
| Propyl | 0.817 | 1.29 |
| $n$-butyl | 0.817 | 1.29 |
| Isobutyl | 0.89 | 0.040 |

TABLE XXIII

$$RBr + C_2H_5O^- \rightarrow C_2H_5OR + Br^-$$

| R | $\Delta\varepsilon^{\neq}{}_\iota$ | Relative speed [7] |
|---|---|---|
| Methyl | 0.74 | 34.4 |
| Ethyl | 0.815 | 1.95 |
| Neopentyl | 0.92 | 0.000008 |

(*loc. cit.*) concerns himself more particularly with the isotope exchange reaction

$$Br^{*-} + R_2 - \underset{\underset{R_3}{\diagup}}{\overset{\overset{R_1}{\diagdown}}{C}} - Br \rightarrow Br^* - \underset{\underset{R_3}{\diagdown}}{\overset{\overset{R_1}{\diagup}}{C}} - R_2 + Br^-.$$

In this case it is reasonable to suppose that in the transition state the two bonds C...Br possess the same length and that the apparent charge of the bromine atoms is a demi-electron. We will therefore agree to say that this complex is 'symmetrical'.

Since it is assumed that the transition state corresponds to a depression on the potential surface, we may estimate the distance C...Br corresponding to this state by finding the one which corresponds to a relative minimum of the energy of a symmetrical intermediate complex.

The passage of the initial molecules to the transition state involves an impact between the $Br^-$ ion and the $CR_1$, $CR_2$ and $CR_3$ bonds. During the impact these bonds are shaken until they become coplanar. Ingold supposes that during this impact the energy of the bonds $CR_1$, $CR_2$ and $CR_3$ vary sufficiently little for one to be able to ignore this variation in the expression of the energy of the complex in terms of the distance $r$ which separates the bromine atom from the central carbon. Under these conditions it is left to us to calculate:

(a) the repulsion between $Br^{*-1/2}$ and the atoms of the groups $CR_1$, $CR_2$ and $CR_3$;

(b) the energy of extension of the bond $C...Br^{-1/2}$.

This latter energy may be estimated by means of a semi-empirical formula of the Morse type

$$\varepsilon_{ext} = D\{e^{-2\beta(r-r_e)} - 2e^{-\beta(r-r_e)}\}.$$

In this formula $D$ represents the dissociation energy of the C—Br bond, $r_e$ signifies its length in the initial molecule and $\beta$ obeys the equation

$$\beta = 0.1227\omega\sqrt{\mu/D}$$

if $\omega$ denotes the frequency of vibration and $\mu$ the reduced mass.

The repulsion between an atom and the $Br^{-1/2}$ ion may be estimated by carrying out the sum of the three following terms:

(a) the energy of polarisation

$$-\frac{\alpha(e/2)^2}{2r^4}$$

(b) the energy of dispersion

$$-\frac{3\alpha\alpha'II'}{2(I+I')r^6}$$

(c) the term connected with the exclusion forces

$$b\cdot e^{-(r/0.345)}.$$

There we can recognise different formulae established in Section 2 of Chapter II. $\alpha$ and $\alpha'$ signify the polarisabilities, $I$ and $I'$ signify the energies of ionisation and $b$ is an empirical coefficient chosen so that the energy obtained by making the sum of the three terms smaller (a), (b), and (c) becomes minimal for an interatomic distance equal to the sum of the Van der Waals' radii of the atoms under study.

The energy of repulsion for a pair of unattached atoms may therefore be finally written

$$\varepsilon_{rep} = -\frac{\alpha(e/2)^2}{r^4} - \frac{3\alpha\alpha'II'}{2(I+I')r^6} + be^{-(r/0.345)}$$

and the fraction of energy of the complex depending greatly on $r$ may be estimated from the formula

$$\varepsilon_{ext} + \sum\varepsilon_{rep}$$

the summation being carried out on all the pairs of unbonded atoms.

Figure 22 shows how this energy varies as a function of $r$. The curve 1 corresponds to the case of methyl bromide and the curve 2 to that of neopentyl bromide. One sees that the minimum of energy is produced for a distance C...Br which does not depend very much on the nature of the alkyl and is between 2.2 and 2.3 Å. Finally Ingold calculated the values of $\varepsilon_{rep}$ for a series of derivatives and for this distance which corresponds to the minimum of energy. Table XXIV contains a few results.

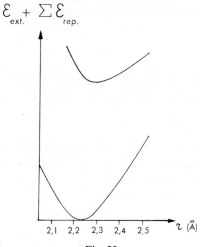

$$\mathcal{E}_{ext.} + \Sigma\,\mathcal{E}_{rep.}$$

| 2,1 | 2,2 | 2,3 | 2,4 | 2,5 |   $\imath$ (Å)

Fig. 22.

TABLE XXIV

| Nature of alkyl | Methyl | Ethyl | Propyl | Isopropyl | Neopentyl |
|---|---|---|---|---|---|
| $\varepsilon$-rep (in kcal mole$^{-1}$) | 0 | 0.9 | 0.9 | 2.3 | 12 |

The measurements of the relative rates have not been carried out in the case of isotopic exchange. One cannot therefore attempt a direct compariosn between these theoretical results and the experimental data. However, one may reasonably think that the qualitative results which are obtained from Table XXIV also concern the reactions described in Tables XXII and XXIII.

It then appears that $\Delta\varepsilon_l^{\neq}$, $\Delta\varepsilon_{rep}$ and the relative speeds vary in the same sense.

$\Sigma\varepsilon_{rep}$ would therefore contribute just like $\Delta\varepsilon_l^{\neq}$ to the slowing down of the reactions when the number of carbon atoms of the alkyls increases. The effect of $\Sigma\varepsilon_{rep}$ becomes spectacular in the case of neopentyl and accounts for the particularly low speed observed in this case.

B. NUCLEOPHILIC SUBSTITUTIONS $S_{N1}$

One frequently designates by the name of nucleophilic substitutions $S_{N1}$ reactions whose first stage consists in a heterolytic dissociation of a bond under the influence of the solvent. The process determining the speed of the reaction is written

$$RX \rightarrow R^+ + X^-$$

when one does not state explicitly the role of the solvent.

In fact it is only from this point that the process is unimolecular. In reality, when one takes the solvent into account, the dissociation is polymolecular and the number of molecules which take part is not well defined.

Dewar [9] thinks that it is possible to represent the intermediate complex by means of the following scheme

$$S \cdots R \cdots X \cdots S$$

in which S denotes a molecule of solvent.

Fukui *et al.* [10] assume that as a consequence the potential barrier will be lower, the greater the charge $q_x$ carried initially by the atom $x$ because the interaction with the solvent will then be more effective. Table XXV makes it possible to compare this charge calculated according to the definition of Sandorfy and Daudel [11] with the relative speed measured experimentally in the case of the hydrolysis of alkyl bromides [12].

TABLE XXV

Electron charges

| Nature of alkyl | $q_x$ | Relative speeds |
|---|---|---|
| Methyl | 1.42 | 1 |
| Ethyl | 1.44 | 1.71 |
| Isopropyl | 1.45 | 44.7 |
| Isobutyl | 1.47 | $10^8$ |

The order of the charges correspond swell with the order of speeds, but it seems probable that other factors such as steric effects also intervene during these reactions. It would in fact be difficult to explain a factor $10^8$ in the reaction speeds by means of a simple variation of charge of a few hundredths of an electron.

C. RADICAL REACTIONS

Many reactions which involve the action of a free radical on a saturated hydrocarbon may be formulated by the equation

$$R-X + Y\cdot \rightarrow [R \ldots X \ldots Y] \rightarrow R\cdot + X-Y.$$

To study the potential barrier of such a reaction, Fukui *et al.* [13] assume that in fact the transition state of the intermediate complex could be represented by the scheme
$$R-X \cdots Y$$

that is to say that the whole of the molecule $R-X$ was not very perturbed in it. Under these conditions, according to Sandorfy and Daudel's method, the secular determinant representing this state of the complex can be written

$$\begin{vmatrix} \alpha_Y - \varepsilon & \gamma\beta & 0 & 0 \\ \gamma\beta & \alpha_X - \varepsilon & \beta_{CX} & \\ 0 & \beta_{CX} & \alpha_C - \varepsilon & \\ 0 & \text{determinant of } R-X & \end{vmatrix}$$

if $\gamma\beta$ denotes the integral of resonance between X and Y.

A calculation of perturbation then enables one to write $\Delta\varepsilon_l^{\neq}$ in the form

$$\Delta\varepsilon_l^{\neq} = \left(\sum_j^{inoc} - \sum_j^{oc}\right) \frac{s_{jx}^2}{\varepsilon_j - \alpha_y} \gamma^2$$

in which $\varepsilon_j$ denotes the energy associated with the $j$-th orbital of RX and $s_{jx}$ signifies the coefficient of the orbital associated with X in the development of this $j$-th orbital. Fukui *et al.* use

$$D_X^Y$$

to denote the coefficient of $\gamma^2$ in $\Delta\varepsilon_l$ expressed in units equal to $-\beta$. Table XXVI shows that in the case of the reaction

$$R-H + R^{\cdot\prime} \rightarrow R^{\cdot} + R^{\prime}-H.$$

$D_H^{R^{\prime}}$ varies with the activation energy measured experimentally [14].

TABLE XXVI

| Hydrocarbon | $D_{R^{\prime}}^H$ | Energy of activation in kcal mole$^{-1}$ when $R^{\cdot} = C_2H_5$[1] |
|---|---|---|
| Methane | 0.99 | 15 |
| $CH_3-CH_2-CH_2-CH_2-H$ | 1.00 | 13.9 |
| $(CH_3)_3C-H$ | 1.03 | 11.6 |

The same observation can be deduced from Table XXVII which relates to the reaction

$$CH_nCl_{4-n} + H^{\cdot} \rightarrow CH_nCl^{\cdot}_{3-n} + HCl.$$

TABLE XXVII

| Chloride | $D_{Cl}^H$ | Energy of activation in kcal mole$^{-1}$ [15] |
|---|---|---|
| $CH_3Cl$ | 1.50 | 7 to 9 |
| $CH_2Cl_2$ | 1.76 | 6 |
| $CCl_4$ | 3.83 | 3.5 |

We will recall what a seat of activation $A$ is. To determine this magnitude one measures the rate constant $k$ of the reaction as a function of the temperature $T$ and then one plots $Lk$ on a graph against $1/T$. As a rule it is not possible to measure $k$ over a wide range of temperature because at low values of $T$ the rates become too slow and at high values of $T$ secondary reactions (particularly dissociations) are superimposed on the reaction studied. Under these conditions the function studied is frequently more or less linear and the product by $\chi$ of the contrary of the slope of the straight line thus obtained is designated by the name activation heat $A$. When it is possible to carry out measurements over a wide range of temperature the function ceases to be linear. The slope which is measured currently therefore represents the derivative of $Lk$ in relation to $1/T$.

The precise definition of the heat of activation is therefore provided by the formula

$$A = -\chi \frac{\partial Lk}{\partial 1/T}$$

from this one deduces

$$A = -\chi \frac{\partial Lk}{\partial T} \frac{\partial T}{\partial 1/T} = T^2 \frac{\partial Lk}{\partial T}.$$

If now one takes into account Formula 1 of Section 1 (Chapter IV) and if we assume

$$f^s = \frac{f_M^s \neq}{f_A^s f_B^s}$$

we will obtain

$$\frac{\partial Lk}{\partial T} = \frac{1}{T} + \frac{\partial L(1 + t^s)}{\partial T} + \frac{\partial L\eta^s}{T} + \frac{\partial L f^s}{\partial T} +$$

$$+ \frac{1}{\chi T^2} \left( \Delta \varepsilon_v^{\neq} + \Delta \varepsilon_l^{\neq} + \Delta \varepsilon_d^{\neq} + \Delta \varepsilon_{n \cdot 1}^{\neq} + \Delta \varepsilon_s^{\neq} (T) \right)$$

The expression of $A$ as a function of molecular magnitudes is therefore written

$$A = \chi T + \chi T^2 \left[ \frac{\partial L(1 + t^s)}{\partial T} + \frac{\partial L\eta^s}{\partial T} + \frac{\partial L f^s}{\partial T} \right] +$$

$$+ \Delta \varepsilon_v^{\neq} + \Delta \varepsilon_l^{\neq} + \Delta \varepsilon_d^{\neq} + \Delta \varepsilon_{n \cdot 1}^{\neq} + \Delta \varepsilon_s^{\neq} (T).$$

It therefore appears that the term $\Delta \varepsilon_l^{\neq}$ which Fukui et al. estimated with the help of $D_X^Y$ is one of the elements of the heat of activation. Tables XXVI and XXVII seem to show that in the case of the reactions studied here $\Delta \varepsilon_l^{\neq}$ constitutes one of the important terms of the heat of activation.

Fukui [16] has also analysed other radical reactions for this same method.

D. CHEMISTRY OF RADIATIONS

The reactions of the removal of hydrogen and chlorine which we have just studied constitute an excellent transition between the field of substitution reactions studied in themselves and certain aspects of the chemistry of radiations.

These reactions are in fact very frequently provoked by irradiations. Thus Yang [17] provoked the reaction

$$H + RH \overset{k_m}{\rightarrow} H^2 + R.$$

(in which RH designates ethane, propane, butane or isobutane) by irradiating these saturated hydrocarbons with $\gamma$ rays. The radiolysis of these hydrocarbons in fact provides the hydrogen atoms necessary for the reaction.

By a process making use of the competitive reaction

$$H + C_3H_6 \rightarrow C_3H_7$$

in which $C_3H_6$ denotes propylene, Yang was able to measure $k_m$ in a large temperature range (from 50 to 250°).

In the above mentioned article the author discusses the results thus obtained by means of the method of linear combinations of bond orbitals of Hall [18] (second book, p. 123). However, he takes greater account of the overlap integrals as was already done by Brown [19] (second book, p. 126).

To represent the intermediate complex Yang bases himself on an idea of Binks and Szwarc [20]. This boils down to assuming that in the intermediate complex:

$$R \cdots H \cdots H$$

the arrival of the new hydrogen nucleus has caused the displacement towards the right of the two electrons of the $R-H$ bond, so that the interaction between the electrons and the rest of the radical R becomes negligible.

It will therefore be assumed that

$$\Delta\varepsilon_l^{\neq} = E[R] - E[R-H] + \text{const}$$

let us state

$$\alpha_{CC} = \int Y_{CH} \, \mathfrak{C} \, Y_{CH} \, dv \qquad\qquad \beta = \int Y_{CH} \, \mathfrak{C} \, Y_{CH'} \, dv$$

$$\beta' = \int Y_{CH} \, \mathfrak{C} \, Y_{CC} \, dv \qquad\qquad S = \int Y_{CH} \, \mathfrak{C} \, Y_{CH'} \, dv$$

$$\gamma = 4(\beta - S\alpha_{CC}) \qquad \text{and} \quad \Delta\alpha = \alpha_C + \alpha_H - 2\alpha_{CC}$$

in which $\alpha_C + \alpha_H$ represents the energy associated with the two electrons of the $R-H$ bond and $\mathfrak{C}$ represents the set operator of the auto-coherent field.

By developing the secular equation according to the powers of S, as Brown did (*loc. cit.*) while only keeping the terms in S and ignoring $(\beta/\beta)^2$ one obtains

$$\Delta\varepsilon_l^{\neq} = \Delta\alpha + N_{CH}S\gamma + \text{const}$$

if $N_{CH}$ denotes the number of CH bonds adjacent to the $R-H$ bond.

It is very reasonable to ignore $(\beta'/\beta)^2$, because Hall (*loc. cit.*) found that

$$\beta'/\beta = 0.2 .$$

If, as previously, the term $\Delta\varepsilon_l^{\neq}$ plays a predominant part in the experimental activation heat, one must expect that this will vary linearly with $N_{CH}$. This is exactly what Yang observed.

Another problem of radiation chemistry which was elegantly dealt with by means of the methods of quantum chemistry is the analysis of the phenomena which take place when saturated hydrocarbons are subjected to the action of a monokinetic electron beam. These substances then undergo dissociations which lead to the formation of positive ions, the nature of which may be determined precisely by means of mass spectrography.

One usually assumes that the first act of the reaction is an impact ionisation of the saturated hydrocarbon. There would thus be first of all formed the monopositive ion of this molecule either in the fundamental electronic state or in an electronically excited state. This ion would then dissociate in fragments.

It was therefore tempting to try to find correlations between the distribution of the positive charge in the ion of the hydrocarbon and the nature of the fragments.

We now know (second book, p. 129) that Hall's method has been applied by Lennard-Jones et al. [21] to the analysis of this charge distribution. The latter have found, for example, that if one numbers the CC bonds starting at one end, the ionisation of the highest level of octane causes the following appearance of population holes:

$$0.035 \text{ in bond } 1, \quad 0.200 \text{ in bond } 3,$$
$$0.115 \text{ in bond } 2, \quad 0.234 \text{ in bond } 4.$$

Generally speaking it appears that in normal hydrocarbons the ionisation corresponding to the *ground state* of the ion is better experienced in the centre of the molecule than towards its extremities.

It therefore seems reasonable to expect that it will be mainly the central bonds of these ions which break at the moment of fragmentation.

Thompson [22, 23] observed that it was in fact the case at least up to octane which supplies mainly fragments of $C_4$ and $C_3$ when it is irradiated with electrons possessing a kinetic energy of 50 eV. Coggeshall [24] observed that this correlation ceases to be verified when one considers longer straight paraffins than octane, because the $C_4$ and $C_3$ fragments remain the most abundant, although the maximum charge which is always situated in the centre of the chain suggests the preponderance of longer fragments.

Lorquet [25] wondered whether the approximations inherent in Hall's method could not constitute one of the causes of this difficulty. Lorquet therefore introduced a few improvements into this method. He took into account all the CH bonds, certain inductive effects and he even went beyond the approximation of the orbitals. The results, however, differ little from those of Lennard-Jones and Hall. Observing that beyond octane the differences between the charges of the CC bonds decrease appreciably, Lorquet concluded that these differences are no longer sufficient to determine the most fragile of the bonds, so that other factors must intervene.

Lorquet (*loc. cit.*) also concerned himself with the study of the case of non-linear paraffins. The distribution of the charges thus obtained in the positive ions of these hydrocarbons suggests that isopentane, 2-methylpentane and 2,3-dimethylbutane should mainly provide $C_3$ ions whilst $C_4$ ions should constitute the principal fragments of 2,2-dimethylbutane and 3-methylpentane. This is in fact what experiments have shown when one uses electrons of 50 eV.

The success of this very simple theory finally seems to a little miraculous. One would normally expect in fact that one would have to develop a more complex theory. We have already said that the electronic impact should sometimes leave a positive ion of the hydrocarbon in an electronically excited state. Everything happens as though such states did not intervene to any appreciable extent. Lorquet thinks that the cause of this apparent simplification of the phenomenon is bound up with the fact that these electronically excited ions fall back very quickly to one of the vibration/rotation levels of the ground electronic state by very rapid non-radiational processes before being dissociated.

To conclude, we will mention some other passages from the literature which deal with similar questions [26, 27].

## 3. A Few Examples of Studies of Rate Constants in the Field of Unsaturated Organic Molecules

### A. SUBSTITUTION ON AN ALTERNANT CONJUGATED HYDROCARBON, LOCALISATION ENERGY AND METHOD OF MOLECULAR DIAGRAMS

Just as in aliphatic chemistry, there exists in aromatic chemistry substitution reactions whose mechanism seems to be fairly well known. We will take as example the nitration of alternant hydrocarbons. Numerous works, particularly those of the teams led by Ingold and by Chédin, have shown that the nitrating agent seems very often to be the nitronium ion $NO_2^+$. When sulphuric acid is used as solvent the nitration very frequently follows a kinetic of order two [28], the rate varying in proportion to the concentration of nitric acid and to that of the substance subjected to nitration. One may therefore simply explain the whole of these facts by assuming that in the presence of sulphuric acid nitric acid is rapidly dissociated according to the reaction

$$NO_3H \rightleftharpoons NO_2^+ + OH^-$$

and that the stage which determines the speed of reaction may be symbolised by the formula

$$NO_2^+ + ArH \rightarrow H^+ + ArNO_2$$

the intermediate complex complying with the formula $ArNO_2H^+$.

As long ago as 1942 Wheland [29], in a famous article, suggested that one should represent the transition state of this complex by assuming that the carbon attacked by the $NO_2^+$ ion behaves like a saturated carbon. The structure of this transition state in the case of the nitration of benzene will therefore be

We will observe that the delocalised bond no longer extends over five carbon atoms and that it is only associated with four electrons because two electrons of the delocalised system of the benzene are now associated with the localised bond joining the carbon attacked to the nitrogen of the nitro group.

According to this model the calculation of $\Delta\varepsilon_d^{\neq}$ will simply be carried out by taking the difference between the energy of this shortened delocalised bond and that of the bond with 6 electrons of benzene. The generalisation to other conjugated hydrocarbons is immediate. It is then usual to designate $\Delta\varepsilon_d^{\neq}$ by the name of *localisation energy*.

In Hückel's approximation (second book, section 29), this energy is written

$$\Delta\varepsilon_d^{\neq} = -2\alpha + k\beta.$$

Wheland thought that this mechanism was of interest for also representing nucleophilic or radical reactions. He was therefore led to use

$$ArH + Y^\varepsilon \;\rightarrow\; Ar\!\!\overset{\displaystyle H}{\underset{\displaystyle Y}{\big\langle}} \;\rightarrow\; ArY + H^\varepsilon$$

to denote in very general terms the essential stage of a substitution in aromatic chemistry, $\varepsilon$ being $+1.0$ or $-1$ according to whether Y is an electrophilic, radical or nucleophilic reagent.

It results from this hypothesis that

$$\Delta\varepsilon_d^{\neq} = -(1 + \varepsilon)\,\alpha + m\beta$$

if one uses Hückel's approximation.

Let us point out that if the substance subjected to nitration is an alternant hydrocarbon, the delocalised bond of the transition state still corresponds to an alternant system but possesses an odd number of centres. The last orbital which may be occupied is therefore non-bonding (second book, p. 170).

The result of this is that the quantity $m$ remains invariable if, without changing the atom attacked of such a hydrocarbon, one passes from the case of an electrophilic reagent to the case of a radical reagent or even to that of a nucleophilic reagent. Then let us compare two substitutions of *the same nature* taking place on two distinct atoms belonging to one and the same alternant hydrocarbon or two different alternant hydrocarbons, and let $\Delta\varepsilon_d^{\neq}$ and $\Delta\varepsilon_d^{\neq\prime}$ be the corresponding energies of localisation, when we will get

$$\Delta\varepsilon_d^{\neq} = -(1 + \varepsilon)\,\alpha + m\beta$$
$$\Delta\varepsilon_d^{\neq\prime} = -(1 + \varepsilon)\,\alpha + m'\beta$$

from which

$$\Delta\varepsilon_d^{\neq\prime} - \Delta\varepsilon_d^{\neq} = (m' - m)\,\beta.$$

Since neither $m$ nor $m'$ depend on the nature of the reagent, *the order in which the*

$\Delta \varepsilon_d^{\neq'} - \Delta \varepsilon_d^{\neq}$ *are classed no longer depends on this and is therefore the same whether the substitution is electrophilic, nucleophilic or radical.*

One must also expect that of all the factors which determine the speed of a substitution on a conjugated hydrocarbon, the localisation energy is one of the most important. One can in fact easily show that except in cases where steric effects intervene the term $\Delta \varepsilon_l^{\neq}$ varies but little when one passes for one and the same type of reaction from attacking a given carbon atom to attacking another carbon atom belonging to the same hydrocarbon or to a different hydrocarbon. This fact is bound up with the fact that $\Delta \varepsilon_l^{\neq}$ only depends on the immediate vicinity of the attacked atom. This vicinity is not very different when one passes, for example, from the nitration of pyrene at 3 to the nitration of pyrene at 2.

The delocalised bond in the transition state corresponding to nitration at 3 of pyrene presents, on the other hand, an overall geometry which is very different from that of the delocalised bond which takes place in the case of nitration of the same hydrocarbon at 2

Hückel's method supplies in fact the following values for the localisation energies of pyrene*

$$\text{at } 2 \quad -(1 + \varepsilon)\,\alpha - 2.27\,\beta$$
$$\text{at } 3 \quad -(1 + \varepsilon)\,\alpha - 2.19\,\beta$$
$$\text{at } 4 \quad -(1 + \varepsilon)\,\alpha - 2.55\,\beta.$$

If one remembers that $\beta$ possesses a negative value it will be seen that it is at 3 that $\Delta \varepsilon_d^{\neq}$ has the lowest value. It is therefore preferably at 3 that the nitration, sulphonation and even halogenation of pyrene takes place [30–32].

Numerous authors have evaluated the energies of localisation by means of more elaborate methods than that of Hückel. Thus K. Fukui et al. [33] used the formalism

---

* Values taken from D.G.T.

of the auto-coherent field with the approximations of Pariser, Parr and Pople and the orbitals of Hückel, but without carrying out the iterations. These authors observed the existence of a more or less linear relationship between the energies of localisation thus calculated and those obtained by means of Hückel's simple method in the case of alternant hydrocarbons. Nesbet [34] showed that this result is preserved when one carries out the entire iterative process normally required by the method of the self-consistent field.

Golebiewski and Sadlej [35] have also shown that from the practical point of view the method of the self-consistent field presents certain advantages. Dewar and Thompson [36] have just used the method of split orbitals with comparable results. Finally let us point out that Chalvet and Daudel [37] already quite a long time ago made use of the still more elaborate method of the interactions of configuration and showed that there exists even there a linear relationship between the results obtained and the energies of localisation calculated by Hückel's method for alternant hydrocarbons.

Another way of improving the calculation of the energies of localisation was introduced by Muller et al. [38]. This consists of taking into account hyperconjugation. Chalvet and Daudel [39] showed that there again there is a more or less linear relationship between the $\Delta\varepsilon_d^{\neq}$ thus calculated and those of Hückel's method. The improvement of Muller et al. comes down therefore to placing

$$\Delta\varepsilon_d^{\neq} = -(1 + \varepsilon)\alpha + \eta m\beta$$

$\eta$ being a coefficient which may depend on the mode of operation. The super-delocalisation introduced by Fukui et al. [40] coincides more or less with the energy of localisation taking hyperconjugation into account. One can therefore see that speaking qualitatively in the field of alternant hydrocarbons one may be content with Hückel's method. If one wishes to obtain an approximate numerical agreement between the relative rates thus obtained when only taking into account $\Delta\varepsilon_d^{\neq}$, that is to say using the very simple formula

$$\frac{k'}{k} = \exp\left[-\frac{\Delta\varepsilon_d^{\neq'} - \Delta\varepsilon_d^{\neq}}{\chi T}\right]$$

with experimental data, it will be sufficient to choose $\beta$ empirically. It was thus that Mason [41] proposed

$$\beta = -7 \text{ kcal} \quad \text{for methylation}$$

and

$$\beta = -16 \text{ kcal} \quad \text{for nitration}$$

If one considers that $\beta$ would have to be constant and that the empirical adjustment does in fact reflect the variation of the coefficient $\eta$ bound up with the hyperconjugation effect, one may conclude with Mason that the intermediate complex of the nitration resembles more Wheland's model than that of methylation.

Contrary to all the methods which we have just examined and which were aimed at improving the calculation of the energy of localisation, certain processes correspond to simplifications of the calculation which is already very simple based on Hückel's method. Thus Dewar has introduced a reactivity index which represents an approximation of the localisation energy calculated within the framework of Hückel's approximation. The reactivity index consists of the energy fraction (expressed in $\beta$ units) due in the term $\Delta\varepsilon_d^{\neq}$ to the variation of energy of the last orbital occupied during the attack. If one uses $c_r$ to represent the coefficients of the orbitals of the carbon atoms adjacent to the atom attacked in the development of this last orbital which may be occupied in the intermediate complex, one can easily show by means of a process of perturbation that the variation of energy corresponding to this orbital can be written

$$\Delta\varepsilon = -\sum_r c_r\beta - \alpha.$$

We have already pointed out that in the case of alternant hydrocarbons which concern us here this orbital is not bonding in the intermediate complex. One therefore has

$$\varepsilon = \alpha.$$

It is therefore not necessary to solve the secular equation to calculate the values of $c_r$. It is sufficient to insert the value $\varepsilon = \alpha$ in the secular system, the solution of which is immediate.

Since usually the last orbital is occupied twice in the initial hydrocarbon, it is the quantity

$$2\Delta\varepsilon$$

which will supply the variation of energy during the attack and the coefficient of $-\beta$ in $2\Delta\varepsilon$, namely

$$2\sum_r c_r$$

and this will be designated by the name *Dewar's reactivity index* [42]. Table XXVIII makes it possible to compare these numbers with the localisation energies.

TABLE XXVIII

| Molecule | Atom attacked | Value of $-m$ in the localisation energy | Reactivity index |
|---|---|---|---|
| Benzene | – | 2.94 | 2.31 |
| Diphenyl | 4 | 2.44 | 2.07 |
| Naphthalene | $\beta$ | 2.40 | 2.12 |
| | $\alpha$ | 2.35 | 1.81 |

The quantitative agreement is not excellent but the order of the reactivity indices coincides more or less with that of the localisation energies. Furthermore Koutecky and Paldus [43] have given theoretical arguments to show that this index may conveniently replace the localisation energy. The example which follows will illustrate how it can be of practical interest. Table XXIX contains information regarding the nitration of phenanthrene. The percentage of each of the isomers which is formed during this reaction was first of all measured by Schmidt and Heinle [44].

Since it is a question of simultaneous reactions, the percentages of the different isomers measure the relative rates of the nitration on the different apices of phe-nanthrene. If therefore $\Delta\varepsilon_d^{\neq}$ were the only important term governing these rates and if the Dewar index was a convenient approximation of this term, one would have to expect that the order of these indices would be the inverse order of the percentages.

Table XXIX shows that if one judges this according to the experimental results of Schmidt and Heinle it is not so at all. Dewar and Warford [45] thought that the origin of the disagreement could be in the errors of the manipulations of Schmidt and Heinle.

TABLE XXIX

| Position of atom attacked | Reactivity index | Percentage of isomer | |
|---|---|---|---|
| | | according to Schmidt and Heinle | according to Dewar and Warford |
| 10 | 1.80 | 60 | 34 |
| 1 | 1.96 | 0 | 27 |
| 3 | 2.04 | 2 | 25 |
| 2 | 2.18 | 20 | 4 |

That is why Dewar and Warford [45] carried out once again the experiments of the German chemists but using much more reliable techniques (chromatography, estima-tion by the method of radio indicators). The corresponding results are shown in the last column of Table XXIX. It seems this time that they agree in an excellent manner with the theoretical predictions.

More generally, the localisation energies appear very often as terms governing the rates of substitution in aromatic chemistry, except where an important steric effect is observed.

Thus Mason [46] showed that the relative speed of bromination of alternant hydro-carbons (Table XXX) is greater, the lower the energy of localisation of the hydro-carbon under consideration, although the mechanism of bromination is doubtless fairly complex [47].

TABLE XXX

| Molecule | Relative speed of bromination at 25° in acetic acid | Coefficient $-m$ of the lowest energy of localisation |
|---|---|---|
| Benzene | 1 | 2.54 |
| Diphenyl | $1.1 \times 10^3$ | 2.4 |
| Naphthalene | $1.4 \times 10^4$ | 2.3 |

Analogous results have been obtained in the case of chlorination. Mason [48] has also shown by involved considerations regarding the intervention of the entropy of solvation that the transition state seems to resemble more Wheland's model, the more reactive the hydrocarbon is.

The *method of molecular diagrams** constitutes another way of studying the reactions which are receiving our attention at the present moment. This method consists in characterising the electronic structure of the molecule under study by a series of indices frequently designated under the name of *static indices* and seeking the relationships between these indices and chemical reactivity. In the case of substitutions on alternant hydrocarbons the important index is known under the name of *free valence number*. This index was introduced under the name of apex charge by Daudel and Pullman [49] in 1945 following an idea of Swartholm [50], using the method of mesomerism. The name free valency index was adopted after a conversation between Coulson and Daudel and is found for the first time in an article by Daudel *et al.* [51]. The notion of the free valency index was introduced by Coulson in 1946 [52] in the method of molecular orbitals. The introduction of this notion constitutes an attempt to render the old idea of residual affinity quantitative.

In order to arrive at this notion one assumes that a given carbon atom can only take part in a limited quantity of bond energy. The sum of the bond indices**, of the bonds starting from this atom, may be considered as a measure of the energy to which the atom contributes. The difference between a certain constant and the sum of these bond indices

$$F_1 = C - \sum_m p_{lm}$$

is a measure of the residual affinity of the atom and constitutes its free valency index. The constant $C$ is arbitrary. It is frequently given the value 1.732. Figure 23 sums up the bond indices and the free valency indices associated with the delocalised bond of naphthalene. One came to think quite naturally that the greater the free valency index of a carbon atom, the easier would become the reactions of substitution on this atom [53]. But naturally it was only a rule based on intuition, the use of which gave results in fairly good agreement with experimental data. It was necessary to wait until 1950 until more solid bases could be found for this rule. Daudel *et al.* [54] then observed in fact the existence of a more or less linear relationship between the free valency index and the energy of localisation by means of the method of states of spin. This important relationship, which constitutes one of the theoretical justifications of the use of static indices for studying chemical reactivity, was discovered by Mrs Roux [55] within the framework of the method of molecular orbitals. Burkitt *et al.* [56] made the same observations when studying the case of other molecules.

---

* The name of molecular diagram was proposed in 1946 by R. Daudel and A. Pullman (*Compt. Rend. Acad. Sci.* **222** (1946) 663) to designate a chemical formula showing the distribution of the static indices along the bonds and near the atoms. This is now very currently adopted.
** See second book, p. 181.

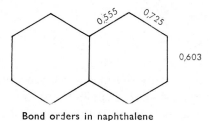

Bond orders in naphthalene

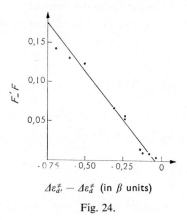

Free valence numbers

Fig. 23.

Figure 24 gives an idea of this relationship between localisation energy and free valence number. We would remind the reader that all that we are saying at this moment only relates to alternating hydrocarbons.

Various authors [57–59] have looked for mathematical reasons for the existence of this relationship. Koutecky *et al.* [60] have also shown by studying a large number of molecules that the relationship of Figure 24 must in fact be represented by three parallel straight lines.

Whatever may be the case, one may with satisfactory approximation replace the localisation energies in the calculation of the rates by free valence numbers. The

$$\Delta\varepsilon_{d'}^{\neq} - \Delta\varepsilon_{d}^{\neq} \text{ (in } \beta \text{ units)}$$

Fig. 24.

formula of p. 129 becomes

$$\frac{k'}{k} = e^{a(F'-F)/\chi T}$$

$a$ representing the proportionality constant which according to Figure 24 makes it possible to pass from the localisation energies to the free valence numbers.

Figure 25 shows, in the case of the methylation of alternant hydrocarbons, the excellent agreement between the experimental results [61] and the results of calculations [62] based on the use of this theoretical formula.

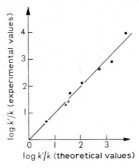

Fig. 25.

We will point out, to conclude this study relating Wheland's model, that Sung *et al.* [63] tried to determine the relative merits of notions such as localisation energy, the reactivity index, free valency index, superdelocalisability by comparing the accuracies which they make it possible to obtain when calculating the relative speeds of reaction. Only the superdelocalisability seems to be slightly less satisfactory than the other three magnitudes from this point of view.

Finally it should be pointed out that certain authors have criticised Wheland's model.

Brown [64] remarked that it was not easy to understand, if one accepts the above-mentioned mechanism, why there is no isotopic effect for the nitration of tritiated derivatives when one exists in the case of sulphonation.

That is why in particular he proposes the following mechanism for the electrophilic substitutions

$$\text{Ar H} + \text{E}^{+} \rightleftharpoons \overset{\text{E}}{\underset{|}{\text{Ar E}-\text{H}}}$$

$$\overset{\text{E}}{+\,\vdots}\;\;\;\;\;\overset{\text{E}}{+\,|}$$

$$\text{Ar}-\text{H} \rightleftharpoons \text{Ar} \ldots \text{H}$$

$$\overset{\text{E}}{+\,|}$$

$$\text{Ar} \ldots \text{H} + \text{B} \rightleftharpoons \text{Ar E} + \text{BH}^{+}$$

According to whether one or other of these different processes is the one which determines the speed of reaction, one may then expect that there is or is not an isotopic effect. In the case of nitration, where there is no isotopic effect, one may assume that it is the first process which determines the speed of reaction. One is then led to the study of the potential barrier which intervenes during the formation of the complex E... $+$ Ar$-$H which Brown considers as a complex by charge transfer. By means of the methods of quantum chemistry Brown has calculated the fraction $\Delta\varepsilon_d^{\neq}$ of this potential barrier and has shown that it does in fact vary like the speed of nitration.

Nagakura and Tanaka [65, 66] conceived of a different mechanism. They assume that the electrophilic reagent first of all removes an electron from the hydrocarbon which becomes a positive ion and that it is on this ion that the reagent which has now become radical fixes itself. They verified that the fixation of the reagent certainly takes place preferably on the apices of the hydrocarbon ion carrying the strongest free valency indices. The interest offered by the mechanism of Nagakura and Tanaka is that of permitting a comparison of various reagents. The first act of the reaction is, in fact, a transfer of electron which will be easier, the higher the last level occupied by the hydrocarbon and the lower the first unoccupied level of the reagent. The first empty level of the $Ag^+$ ion is, for example, much higher ($-7.6$ eV) than that of the $NO_2^+$ ion ($-11.0$ eV). One would therefore understand why the $NO_2^+$ ion is an active agent for substitution whilst the $Ag^+$ ion restricts itself to the formation of complexes.

The last occupied level of benzene ($-3.74$ eV) is distinctly higher than that of methane ($-13$ eV), which would explain why it is easier to nitrate benzene than methane.

It would be premature to make a definitive choice between these different mechanisms, but pure chemists will doubtless appreciate the efforts of quantum chemists who are trying construct theories which take ever better account of the whole of the experimental facts known in a given field and in this way facilitate the understanding of the quantitative role of each of the numerous factors which determine chemical reactivity.

B. OTHER EXAMPLES OF SUBSTITUTIONS ON AN ATOM BELONGING TO THE DELOCALISED
    SYSTEM OF A CONJUGATED MOLECULE

As soon as one leaves the field of alternant hydrocarbons, the study of the substitution reactions becomes much more complex. The highest orbital which may be occupied in the transition state of Wheland ceases to be non-bonding so that the coefficient $m$ of the localisation energy of a given atom depends on the character of the reagent. Table XXXI shows this fact in the case of pyridine. The calculations were carried out with the approximations of Hückel and the parameters chosen by Yvan [67].

One can therefore see that the effect of the term $\Delta\varepsilon_d^{\neq}$ no longer orientates the reaction in a centre which is independent of the character of the reagent. In the case of pyridine, $\Delta\varepsilon_d^{\neq}$, according to these calculations, orientates a radical reaction to ortho, an electrophilic reaction to meta and a nucleophilic reaction to ortho.

Another complication arises from the multiplicity of possible choices of parameters.

A third complication is bound up with the fact that, even when the problem of the

TABLE XXXI

Localisation energies

| Reagent | Position of carbon | | |
|---|---|---|---|
| | 2 | 3 | 4 |
| radical | $-\alpha - 2.45\beta$ | $-\alpha - 2.52\beta$ | $-\alpha - 2.54\beta$ |
| electrophilic | $-2\alpha - 2.73\beta$ | $-2\alpha - 2.52\beta$ | $-2\alpha - 2.85\beta$ |
| nucleophilic | $-2.16\beta$ | $-2.52\beta$ | $-2.22\beta$ |

choice of parameters does not arise, the orders of the energies of localisation may depend on the method of calculation. Thus Fukui *et al.* [68] observed that the order of the energies of localisation calculated by Hückel's method differs from that obtained by means of the self-consistent field method with the approximations of Pariser and Parr. Chalvet *et al.* [69] recently confirmed this fact by examining a larger number of examples. Fortunately these authors reached the conclusion that the atom designated as having to be *the most* reactive in relation to a given reagent remains independent of the method followed to calculate the energy of localisation. It is not the same for the position of the secondary centres of reactivity.

The use of static indices also enjoys a special character. As is known, contrary to what happens in the case of alternant hydrocarbons, the electron charge associated with the delocalised bond is distributed unequally between the different atoms (second book, Section 30 and Sections 40, 41, 42). One was therefore very quickly tempted to assume that the most negative atoms should be the seats of electrophilic reactions and the most positive atoms should be those of nucleophilic reactions [70]. The free valency index was only evoked for interpreting the orientation of radical reactions [71]. Furthermore it was very soon perceived [72, 73] that the centres designated by the static indices do not always coincide with those which are suggested by the consideration of the localisation energies.

A great deal of caution is therefore called for in this field and we think that the study of a series of particular cases is the best method for testing out the climate which characterises this region of quantum chemistry.

We will envisage first of all the case of non-alternant hydrocarbons. This raises a basic problem because, as we have already pointed out (Subsection 3D of Chapter II), Dewar and Gleicher are of the opinion that the classic treatments exaggerate considerably the delocalisation in non-alternant hydrocarbons, with the exception of azulene. However, the localisation energies calculated by Chalvet *et al.* [74] by the method of the self-consistent field indicate very correctly the most reactive centres in relation to electrophilic, nucleophilic and radical reagents for many of these hydrocarbons, namely: position 5 for the nitration of azulene [75], position 3 for the attack of this same molecule by the nucleophilic reagent LiCH$_3$ [76], and again the position 5 for the action of the phenyl radical [77], and finally position 1 for the nitration of acenaphthylene [75], that of fluoranthene [75] and position 1 for nucleophilic substitutions in fulvene [78].

fulvene          azulene

acenaphtylene

fluoranthene

We would also point out that the calculations of Chalvet *et al.* indicate the existence of an exceptionally low electrophilic localisation energy in the case of pentatriafulvalene. This hydrocarbon does not seem to have been studied experimentally but Kende [79] *et al.* have just shown in a close derivative that there is a strong aptitude to give reactions of electrophilic substitution.

The study of the aza derivatives of alternant hydrocarbons has formed the subject of a very large number of works. Pullman and Effinger [80] have shown that the free valency indices just like the radical localisation energies take quantitatively into account the relative speeds of methylation by the $CH_3$ radical, in a series of molecules such as pyridine, quinolein, isoquinolein.

We have seen (Table XXXI) that the lowest nucleophilic localisation energy is found at 2 in the case of pyridine. In the presence of potassium amidide pyridine gives in effect 2-aminopyridine with a yield of 82 to 90% [81]. Nasielki, however, has pointed out to me [82] that it is perhaps in fact the loss of the $H^-$ ion by the intermediate complex which would be the determining stage of these reaction speeds because the addition of oxidising agents accelerates the reaction [83]. In the case of other nucleophilic reactions such as reduction, hydroxylation, alkylation, cyanation, the great uncertainties which exist regarding the mechanism [81] render discussions somewhat risky.

Table XXXI suggests that the electrophilic reactions should take place at 3. It is in fact there that nitration [84], halogenation [85] and sulphonation [86] take place.

In fact one must not forget that as these reactions take place in a very acid medium it is the pyridinium ion which reacts; one will therefore have to think regarding the localisation energies of this ion. But, since it suffices to increase the parameter $\alpha_N$ of the nitrogen a little to take into account the presence of the proton, one obtains energies of the same order as those of Table XXXI. This means to say that for qualitative reasonings one can discuss on this basis.

The case of pyrimidine and that of pyrazine have been examined by Brown and Hefferman [87] and that of pyridine oxide by Barnes [88].

Brown and Harcourt [89] have just discussed in very great detail the case of quinolein. They have even used the theoretical results for *proposing mechanisms of reaction.* It would take up too much space here to summarise the whole of the results reported by these authors and which we find extremely interesting. We will restrict ourselves to reporting a few observations which seem to us to be of very particular interest.

It is a question in particular of reactions regarded as nucleophilic substitutions such as the action of $R^-$, $OH^-$, $CN^-$ ions on a salt of 1-alkylquinolinium. With the $R^-$ and $OH^-$ ions, which are regarded as very energetic reagents, the reaction takes place at 2 whereas with an ion such as $CN^-$, which behaves as a mild nucleophilic agent, the reaction takes place at 4. Now with reasonable parameters it is possible to find a stronger positive charge at 2 than at 4 during a calculation based on Hückel's approximation, whilst the localisation energy remains lowest at 4.

*Brown and Harcourt draw the conclusion from this that it is the charge which should determine the speeds in the case of energetic reagents where the potential barrier remains weak and the localisation energy in other cases.*

The same authors also try to understand why according to experimental conditions the nitration of quinolein may take place preferably at 3, 5 and 6 *or* at 8.

In order to take into account the results of nitration in an acetic medium they were led to suggest the following mechanism

Without wishing to consider the establishment of this mechanism as being very convincing, this example nevertheless shows how a theoretical analysis of a problem of reactivity may assist in suggesting a plausible mechanism. *May a reaction mechanism be better than plausible?* We will note incidentally that Austin and Ridd [90] have just examined afresh in a very quantitative manner the nitration of quinolein. They arrive at the conclusion that in this field the calculation of the localisation energies alone is insufficient for interpreting the reactivity. This is doubtless a conclusion which one must bear in mind in general terms in heterocyclic chemistry [91, 92].

Brown and Harcourt discussed with the same care the case of isoquinolein [93]. The choice of parameters was in particular the subject of an excellent analysis.

Kwiatkowski and Zurawski [94] examined in detail the group 'pyridine, pyrimidine and pyridazine'. They observe that in this case the static indices and the dynamic indices agree with one another on the whole and permit of the interpretation of the majority of reactions, particularly when looking for the most active centres inside one and the same molecule.

But the most complete study of the reactivity of the aza derivatives of alternant hydrocarbons by the methods of quantum chemistry is without doubt that of Zahradnik and Parkanyi [95].

It deals in fact with 17 compounds; the static indices and the dynamic indices have all been calculated. From the whole of the results and discussions it would appear that by assuming that the charges represent the valid index for energetic reagents and localisation energies are an adequate criterion for mild reagents, the majority of reactions appear to be satisfactorily interpreted. However, serious difficulties remain in certain cases where no index seems to suffice.

We will mention analogous studies on furane, pyrole, benzofurane, dibenzofurane and carbazole [96] and on pyrazole and imidazole [97, 98] thiophene [99], 1-3-thiazole and the thiadiazoles [100].

The substitution reactions on the substituted compounds of conjugated hydrocarbons very soon became the subject of interesting studies [101, 102], whether the substituents are electron donors such as $NH_2$, OH, F, Cl, Br, I or electron acceptors such as $NO_2$. The study of the properties of aniline would deserve to be taken up again particularly in the light the experimental works of Brickman and Ridd [103]. A few indications regarding the phenols are to be found in an article by Igea Lopez-Vazquez [104]. The problem of quinones was broached by Pullman and Diner [105] as well as by Senent and Igea [106].

The reactions of nucleophilic substitutions have just formed the subject of interesting quantitative works. Murto [107] has for example found a more or less linear relationship between the charge on the carbon attacked and the logarithm of the speed of methoxylation of nitrated derivatives.

Simonetta and Carra [108] have found a linear relationship between the activation energy of nucleophilic reactions carried out on nitrated derivatives and the term $\Delta \varepsilon_d^{\neq}$ calculated with a model of the transition state which generalises for this case the model of Muller, Pickett and Mulliken. Finally we would observe that L. B. Kier [109] has found an excellent relationship between the localisation energies and the speed of hydrolysis of molecules containing the carbonyl group such as urea, acetamide, urethane, methyl acetate etc.

C. OTHER TYPES OF SUBSTITUTION

To terminate this long section devoted to substitutions, it remains for us to describe a few studies which are rather particular. Fierens *et al.* [110] measured with very great care the speed of exchange between sodium iodide and the chloromethylated derivatives of alternant hydrocarbons in solution in acetone.

In the case of the benzene derivative the reaction is written

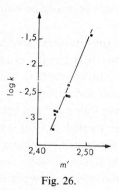

The mechanism of the reaction doubtless resembles that of Walden's inversion. One may therefore assume that the transition state would be well represented by the following formula

In order to calculate the term $\Delta\varepsilon_d^{\neq}$ corresponding to this mechanism, Daudel and Chalvet [111] used a technique which recalls that of hyperconjugation. Let $\chi_1$ and $\chi_2$ be the orbitals which can be associated with the C...Cl and C...I bonds of the complex. One may form the two combinations $\chi_1 - \chi_2$ and $\chi_1 + \chi_2$. The first is pseudo-antisymmetric vis-à-vis the benzene plane and for this reason Daudel and Chalvet introduced it into the delocalised system. The calculation of $\Delta\varepsilon_d^{\neq}$ was then carried out according to an approximation of the Hückel type and Figure 26 shows the remarkable linear relationship which connects the coefficient $m'$ of $\beta$ in $\Delta\varepsilon_d^{\neq}$ and the logarithm of the rate constant measured by the Belgian chemists.

Fig. 26.

## D. THE DIENE SYNTHESIS AND ADDITION REACTIONS

From their very first work on the method of molecular diagrams, Daudel and Pullman [112] concerned themselves with the diene synthesis, nothing that the dienes possess at their ends atoms which carry a powerful free valency index and that the philodienes are usually bonds formed by two atoms which also possess considerable free valency indices.

In the light of the study of this Diels Alder reaction, the generalisation of the notion of localisation energy was carried out by Brown [113]. As always, the problem of the mechanism of the reaction is posed from the very beginning. It has been analysed by numerous authors [114]–[119]. Two mechanisms appear probable. According to one of these the fixation of the philodiene on the diene is supposed to take place in one stage. The diagram which follows gives an example of diene synthesis carried out by this mechanism

It is clear that if one applies Wheland's ideas here and generalises them, the transition state of the intermediate complex will not differ a great deal from the final product. The term $\Delta\varepsilon_d^{\neq}$ in the case of the addition of maleic anhydride to anthracene is reduced, for example, to the simple difference between twice the energy of the delocalised system of benzene and the energy of that of anthracene. This energy $\Delta\varepsilon_d^{\neq}$ which then results from the appearance of localised bonds on the two apices situated in the para position is called the *para-localisation energy*. Although one lacks the kinetic measurements of quality, Brown has been able to show that the lower this energy is, the easier is the Diels–Alder reaction.

In the case of alternant hydrocarbons, the reaction only seems to take place if one of the pairs of atoms situated in the para position in the molecule possesses a localisation energy lower than $3.6\beta$ in the case of maleic anhydride and classic experimental conditions [120]. Brown [121] has also studied the effects of substituents and hetero atoms on the para-localisation energy.

Chalvet *et al.* [122] as well as Brown [123] have also shown that there exists a decreasing linear relationship between the sum of the free valency indices of the atoms attacked and the para-localisation energy, thus establishing a theoretical basis for the purpose of the use of the method of molecular diagrams for the study of diene synthesis. We will finally note that Fernandez *et al.* [124] observed that the method of the free electron (second book, Section 33C) also gives good results.

The second plausible mechanism boils down to supposing that the reaction takes place in two stages. It has not been used systematically for quantum calculations.

However, one should note that the para-localisation energy does not enable one to take into account certain number of characteristics of the diene synthesis. Let us consider the fixation of an asymmetrical dienophile on an also asymmetrical diene, as for example the diene syntheses set out in the following formula

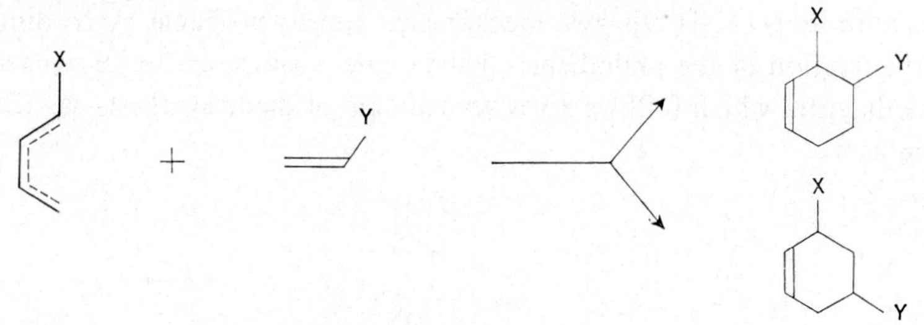

In certain cases a single isomer is formed, although the para-localisation energy remains the same in both cases.

To explain this phenomenon one may make use of the transition state suggested by Evans [125]. The following formula corresponds to this hypothesis in the case of the fixation of acrolein on to 1-phenylbutadiene

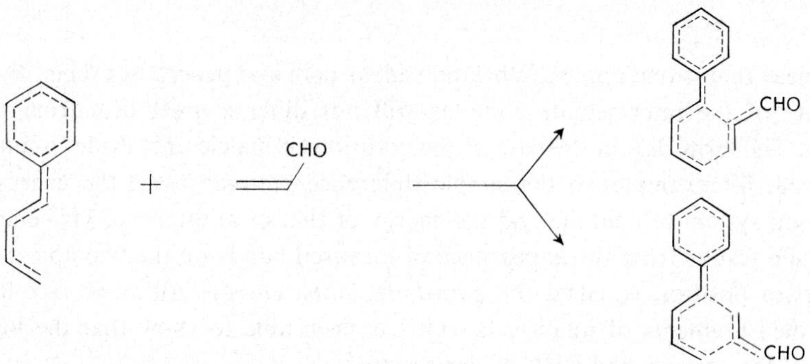

It can be seen here that the delocalised bond is supposed to extend to the acrolein in the transition state. Consequently its energy is not the same in both cases. Streitwieser [126] observed that if the aldehyde group is directly conjugated with the phenyl, one must expect a weaker potential barrier, which explains why it is the substance

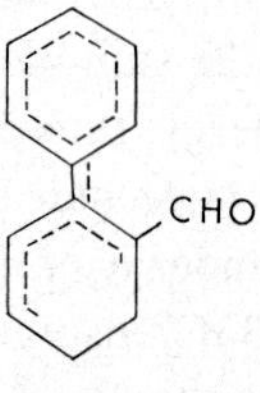

which is preferably formed during the reaction [127]. The use of the same type of transition state explains why butadiene is better fixed than ethylene on to dienes (Streitwieser, *loc. cit.*).

Numerous other addition reactions have formed the subject of studies by means of similar methods. In the case of reactions of fixation on a bond of a conjugated system of reagents such as osmium tetroxide, ozone, the ortho-localisation energy replaces the para-localisation energy and the bond index plays the part of the sum of the free valence numbers [128].

The case of reagents which appear to dissociate prior to addition has also been discussed by Daudel *et al.* in the book *Quantum Chemistry*. We would also refer to the interesting study by Pilar [129] concerning the fixation of such a reagent either at 1,2 or at 1,4 on to butadiene.

## E. OTHER REACTIONS

In order to show the variety of subjects which one can deal with by means of the methods of quantum chemistry, we will refer to the researches of Yang [130] on the fixation of a hydrogen atom on to mono-olefins, butadiene, benzene, which are of interest to radiation chemistry, the work of Fueno *et al.* [131] on the anti-oxidant action of phenols, the analysis of the phenomena of reduction [132, 133] of rearrangements of Fritsch [134], of Claisen and Cope [135], of the reactivity of organo-metallic compounds [136], without mentioning researches on heterogenic catalysis, and certain electrochemical phenomena which fall outside the framework of our work which is reserved to the study of phenomena in a homogeneous phase.

## F. POLYMERISATION REACTIONS

We will conclude these studies devoted to thermochemical reactions by analysing a few recent works devoted to the important field of polymerisation reactions. We shall restrict ourselves to the treatment of radical polymerisations, referring the reader for other types of polymerisation to the book of Higasi *et al.* [137].

Let us therefore consider the reaction of initiation of a polymerisation chain making

use of the radical $\begin{matrix} H \\ -\,C \\ Y \end{matrix}$ and the monomer $\begin{matrix} H \quad\quad H \\ \diagdown \quad \diagup \\ C{=}C \\ \diagup \quad \diagdown \\ H \quad\quad X \end{matrix}$ . Yonezawa *et al.* [138] as-

sumed that the transition state corresponding to this reaction could be written

$$-\underset{\underset{Y}{\|}}{\overset{\overset{H}{|}}{C}}\cdots\underset{\underset{H}{|}}{\overset{\overset{H}{|}}{C}}{=}C\overset{\diagup H}{\underset{\diagdown X}{}}$$

They calculated the energy stabilisation $\Delta\varepsilon_d^{\neq}$ which results from the extension of the delocalised system according to the process which Dewar used to calculate his reactivity indices. Table XXXII shows that there exists a very convenient relationship between these values of $\Delta\varepsilon_d^{\neq}$ and the relative rates of chain initiation.

The authors have also observed that the polymerisation only starts when the value of $\Delta\varepsilon_d^{\neq}$ exceeds 0.83.

TABLE XXXII

| Nature of monomer | Relative speed | $\Delta\varepsilon^{\neq}{}_d$ (in $\beta$ units) |
|---|---|---|
| Acrylonitrile | 2.44 | 1.18 |
| Styrolene | 1 | 0.93 |
| Vinyl chloride | 0.06 | 0.85 |

Fueno *et al.* [137] observed on the other hand that the rate of initiation varies with the localisation energy of the carbon attacked in the monomer in cases where the initiation is produced by a methyl radical (Table XXXIII).

TABLE XXXIII

| Nature of monomer | Relative speed | Localisation energies (in $-\beta$ units) |
|---|---|---|
| Acrylonitile | 1540 | 1.69 |
| Styrolene | 820 | 1.704 |
| Vinyl acetate | 37 | 1.88 |
| Propylene | 22 | 2.01 |

Finally it is now a question of studying the speed of propagation of the chain

$$
\begin{array}{ccc}
& H & \\
R-CH_2-C^{\cdot} + CH_2 = CH & \rightarrow & R-CH_2-CH-CH_2-\overset{\cdot}{C}H \\
X & & X \qquad\quad X \\
& X &
\end{array}
$$

and one will see that it is the localisation energy of the reactive carbon of the *free radical* which seems to determine the speed of the reaction, which enables us to understand why the speed of initiation and the speed of propagation frequently vary in opposite directions (Table XXXIV).

TABLE XXXIV

| | Relative speed at 60° | Localisation energies in the radical (in $-\beta$ units) |
|---|---|---|
| Vinyl acetate | 2300 | 0.445 |
| Methyl methacrylate | 705 | 0.691 |
| Styrolene | 145 | 0.721 |

## 4. The Reactivity of Organic Molecules in an Excited Electronic State

A. INTRODUCTION

It seems that A. Pullman and R. Daudel were the first to concern themselves with the electronic distribution of a conjugated molecule placed in an excited electronic state. From this point of view they studied [140] the delocalised butadiene bond according to the method of molecular diagrams and observed that, whilst in the ground state the said bond has strong bond indices at 1-2 and 3-4 and a weak bond index at 2-3, in the first excited electronic state a strong bond index appears at 2-3 to the detriment of the indices at 1-2 and 2-4, which become weak. The result of this is that in this state the free valency indices at 1 and 4 take up very strong values.

Since the above-mentioned authors used the method of mesomerism one may express these results in other terms. The electronic wave function may be described symbolically

$$\Psi = a \;\diagdown\!\!\diagup\!\!\diagup + b \;\diagdown\!\!=\!\!\diagup$$

in which naturally it is well to understand that the formulae denote the electronic functions which the mesomerism method permits one to associate with them. In the ground state $a$ is large and $b$ is small. In other words, the butadiene bond resembles Kékulé's formula. In the first excited state the contrary is the case, the butadiene must resemble Dewar's formula. This observation suggests that under the influence of light one could expect a cyclisation of butadiene.

$$\diagdown\!\!\diagup\!\!\diagup + h\nu \to \diagdown\!\!=\!\!\diagup$$

In 1946 no experimental data was known in favour of such a reaction but in 1963, Srinivasan [141] achieved this by irradiating butadiene in dilute ether solution.

In 1948 Crawford and Coulson studied by the same method the photodimerisation of acenaphthylene [142].

In 1951 Fernandez Alonso whilst staying in our laboratory studied the distribution of electron charges of mononitrobenzene in the ground state and in the first excited electronic states. Figure 27 shows the results which he obtained for the ground state and one of these excited states. It can clearly be seen that in this excited state the orientation effect of the meta group is very different from what happens in the ground state. In the latter case, the nitro group does in fact impoverish the *ortho* and

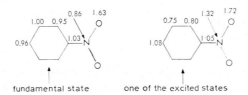

fundamental state          one of the excited states

Fig. 27.   Distribution of electronic charges in nitrobenzene.

*para* apices of the cycle, whilst in the excited state the impoverishment is effected mainly at *meta*.

A little later, Fernandez and Domingo [143] obtained analogous conclusions in the case of nitro-anilines. At that time one did not know any experimental facts which could be confronted with these theoretical results.

But in 1956 Havinga *et al.* [144] studied the photochemical hydrolysis of the sulphates and phosphates of nitro-phenols and observed that the *meta* derivative hydrolyses much more quickly. One may interpret this phenomenon by assuming that, in accordance with theory, the nitro group in an excited state is capable of impoverishing the electronic charges at *meta*, thus facilitating the heterolytic cuts in this region.

A little later Zimmermann [145] found analogous results in the case of the photochemical solvolysis of the trityl ethers of nitrophenols.

The whole of these first observations therefore lead one to hope for the possibility of fruitfully using the method of molecular diagrams in the study of photochemical reactions.

Nevertheless, there are many reasons which seem to oppose the use of this method in this case.

One of the first difficulties which is met with when one tries to study photochemical reactions theoretically resides in the fact that the experimental data generally take the form of quantum yields and, as Porter pointed out [146], it is a long way from this yield to the speed of reaction.

Let us consider in fact a typical photochemical reaction: a molecule A passes from its ground state to an excited state A*. Let us use $k_a$ to designate the ordinary deactivation speed constant of this state (including in particular deactivations by impact). Let us suppose, as is frequently the case, that by internal conversion it is possible to reach a level A* at which the reaction under study is produced. Let $k_b$ be the speed constant of this conversion. This state could deactivate itself with a speed constant $k_c$. Finally we will use $k_R$ to denote the speed constant of the reaction (Figure 28).

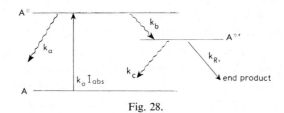

Fig. 28.

It is easy to see that the quantum yield $\gamma$ is written

$$\gamma = \frac{k_R k_b}{(k_a + k_b)(k_c + k_R)}.$$

It can clearly be seen from this formula that modifications of $\gamma$ may be connected with variations in speeds of deactivation or internal conversion and not faithfully reflect the variations of $k_R$, the real chemical reactivity of the state under study.

In principle one should therefore not content oneself with quantum yields but also measure $k_a$, $k_b$, $k_c$ to reach $k_R$. Unfortunately these operations are tricky and the cases are very rare for which these measurements have been carried out.

In the absence of anything better one is therefore compelled to *assume that γ varies with* $k_R$ but one must not forget that this is a hypothesis which may well be unfounded.

We have seen that on the other hand the use of static indices does not seem clearly justified in the case of the chemistry of the ground state except by reason of their relationship with the dynamic indices bound up with the theory of the transition state. Now, from the theoretical point of view, as Laidler [147] remarked, it seems difficult to see to what extent the theory of the transition state remains valid for reactions which take place from excited states.

Furthermore, the temperature coefficients measured on the quantum yield are frequently very small or even negative [148, 149]. It is therefore probable that the potential barriers, if they exist, are very low, an observation which is well in agreement with the fact that the photochemical rate constants are very high.

It therefore seems essential to take up the question again in detail so as to judge if the few successes presented in the introduction correspond to fortuitous coincidences or conceal true correlations.

## B. ELECTROPHILIC SUBSTITUTIONS

De Bie and Havinga [150] have studied in very great detail the photochemical deuteration of nitrobenzene, anisol and the nitro-anisols. They assume that this reaction behaves like an electrophilic substitution and consequently they have calculated the charges and electrophilic localisation energies both for the ground state and for the first excited electronic state.

Table XXXV summarises certain of their results.

It will be seen that the charges and localisation energies account just as happily for the reactivity of the ground states. On the other hand, in the case of the excited

### TABLE XXXV

| Rates | | Charges | Localisation energy |
|---|---|---|---|
| Anisol (ground state) | ortho 0.067 | 1.08 | 0.342 |
| Anisol (ground state) | meta nil | 0.98 | 0.500 |
| Anisol (ground state) | para 0.052 | 1.04 | 0.353 |
| Anisol (excited state) | ortho 7.5 % in 4 h | 1.09 | 0.078 |
| Anisol (excited state) | meta 8 % in 4 h | 1.09 | 0.136 |
| Anisol (excited state) | para 1 % in 4 h | 0.95 | 0.182 |
| Nitrobenzene (ground state) | ortho nil | 0.96 | 0.634 |
| Nitrobenzene (ground state) | meta nil | 1 | 0.632 |
| Nitrobenzene (ground state) | para nil | 0.96 | 0.653 |
| Nitrobenzene (excited state) | ortho 1 % in 4 h | 0.91 | 0.180 |
| Nitrobenzene (excited state) | meta 0.5 % in 4h | 0.89 | 0.161 |
| Nitrobenzene (excited state) | para 8.6 % in 4h | 0.99 | 0.213 |

states, it is only the charges which are suitable; the localisation energies give deceptive results. Feler [151] made the same observation in the case of the nitro-anisols.

The situation therefore has something comparable with what we have met when presenting the reactions studied by Brown and Harcourt in the case of quinolein or reactions involving energetic reagents (therefore rapid reactions) which rather followed the charges than the localisation energies. It would certainly be very useful to analyse this problem in greater depth.

### C. RETURN TO THE CASE OF BUTADIENE

We pointed out in the introduction that for a short time one has known how to carry out the photocyclisation of butadiene and we have interpreted the reaction as if this cyclisation took place in the first excited electronic state. In fact, as Dauben stated [152], one should ask oneself whether the reaction could not be produced from a level of vibration of a high quantum number corresponding to the fundamental state.

One may invoke Woodward and Hoffman's rule [153] to attempt a choice between these two mechanisms. These authors assume that the orbitals $\pi$ corresponding to the last orbital occupied must at the moment of cyclisation turn so that their parts of the same sign overlap one another.

Let us consider a trans-trans disubstituted derivative of butadiene. Since in the fundamental state the highest orbital occupied is antisymmetrical in relation to the

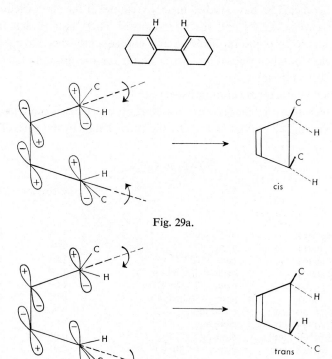

Fig. 29a.

Fig. 29b.

median plane of the central bond, this rule suggests the formation of a trans derivative (Figure 29a). In the case of the first excited state, the highest orbital occupied is symmetrical in relation to this plane and must lead to a cis derivative (Figure 29b). The only derivative of butadiene for which the steric nature of the product formed is known is 1,1-bicyclohexenyl.

The photocyclisation of this substance leads to a cis derivative. There are therefore serious chances that photo-cyclisation will take place here from the electronically excited state. We will point out that by analogous means Longuet-Higgins and Abrahamson [154] have analysed the conversion of cyclobutene into butadiene.

D. MISCELLANEOUS STUDIES

We would point out to conclude this section that the method of molecular diagrams has been applied to the study of widely differing photochemical reactions such as the photo-oxidation and photodimerisation of acenes, the photo-isomerisation of stilbene, the photochrome properties of pyranospiranes and the photodimerisation of thymine.

## References

[1] E. D. Hughes, *Quart. Rev.* **5** (1951) 245.
[2] On this subject see: C. K. Ingold, *Les Réactions de la Chimie Organique*, Hermann et Cie, Paris, 1948, p. 27.
[3] K. Fukui, H. Kato and T. Yonezawa, *Bull. Chem. Soc. Japan* **33** (1960) 1201.
[4] C. Sandorfy and R. Daudel, *Compt. Rend. Acad. Sci.* **238** (1954) 93.
[5] F. Fukui in *Modern Quantum Chemistry*, Ac. Press, New York, 1965, p. 49.
[6] C. K. Ingold, *Structure and Mechanism in Inorganic Chemistry*, Cornell University Press, Ithaca, N.Y., 1953.
[7] C. K. Ingold, *Structure and Mechanism in Organic Chemistry*, Cornell University Press, Ithaca, N.Y., 1953.
[8] C. K. Ingold, *Les Réactions de la Chimie Organique*, Hermann et Cie, Paris, 1948, p. 31.
[9] M. J. S. Dewar, *Electronic Theory of Organic Chemistry*, Clarendon Press, Oxford, 1949, p. 64.
[10] K. Fukui, H. Kato, and T. Yonezawa, *Bull. Chem. Soc. Japan* **33** (1960) 1201.
[11] C. Sandorfy and R. Daudel, *Compt. Rend. Acad. Sci.* **238** (1954) 93.
[12] C. K. Ingold, *Structure and Mechanism in Organic Chemistry*, Cornell Univ. Press, Ithaca, N.Y., 1953.
[13] K. Fukui, H. Kato and T. Yonezawa, *Bull. Chem. Soc. Japan* **34** (1961) 111.
[14] N. N. Tikhomirova and V. V. Voevodski, *Chem. Abstr.* **45** (1951) 9940.
[15] E. W. R. Steacie, *Atomic and Free Radical Reactions*, Reinhold Co., New York 1946; *ibid.*, 2nd ed., Vol. II 1954.
[16] K. Fukui, *Modern Quantum Chemistry*, Part I, Ac. Press, New York, 1965, p. 49.
[17] K. Yang, *J. Phys. Chem.* **67** (1963) 562.
[18] G. G. Hall, *Proc. Roy. Soc.* A **205** (1951) 541.
[19] R. D. Brown, *J. Chem. Soc.* (1953) 2615.
[20] J. H. Binks and M. Szwarc, *J. Chem. Phys.* **30** (1959) 1494.
[21] J. Lennard-Jones and G. G. Hall, *Trans. Far. Soc.* **48** (1952) 581.
[22] R. Thompson, *Conference on Applied Mass Spectroscopy*, Institute of Petroleum, ed., London 1953, p. 185.
[23] See also: F. H. Field and J. L. Franklin, *Electron Impact Phenomenon*, Ac. Press, New York, 1957, p. 171.
[24] N. D. Coggeshall, *J. Chem. Phys.* **30** (1959) 595.
[25] J. C. Lorquet, *Molec. Phys.* **9** (1965) 101.
[26] K. Fukui, H. Kato, and T. Yonezawa, *Bull. Chem. Soc. Japan* **35** (1962) 1475.

[27] K. Fukui, *Modern Quantum Chemistry*, Vol. I, Ac. Press, New York, 1965, p. 49.
[28] On all these questions see: C. K. Ingold, *Les Réactions de la Chimie Organique*, Hermann, et Cie, Paris, 1948, p. 44.
[29] G. W. Wheland, *J. Am. Chem. Soc.* **64** (1942) 900.
[30] H. Vollman, H. Becker, M. Corell, H. Streeck, and G. Langbein, *Ann.* **531** (1937) 1.
[31] G. Lock, *Ber.* **70** (1937) 326.
[32] N. P. Buu-Hoi and J. Lecocq, *Compt. Rend. Acad. Sci.* **226** (1948) 87.
[33] K. Fukui, K. Morokuma, and T. Yonezawa, *Bull. Chem. Soc. Japan* **32** (1959) 1015.
[34] R. K. Nesbet, *J. Chim. Phys.* **59** (1962) 754.
[35] A. Golebiewski and A. J. Sadlej, *Bull. Ac. Polon. Sci.* **13** (1965) 735.
[36] M. J. S. Dewar and C. C. Thompson, in press.
[37] O. Chalvet and R. Daudel, *J. Chim. Phys.* **49** (1952) 629.
[38] N. Muller, L. W. Pickett, and R. S. Mulliken, *J. Am. Chem. Soc.* **76b** (1954) 4770.
[39] O. Chalvet and R. Daudel, *Compt. Rend. Acad. Sci.* **241** (1955) 413.
[40] K. Fukui, T. Yonezawa, and C. Nagata, *Bull. Chem. Soc. Japan* **27** (1954) 423.
[41] S. F. Mason, *J. Chem. Soc.* (1958) 4329.
[42] M. J. S. Dewar, *J. Am. Chem. Soc.* **74** (1952) 3357; *Prog. in Org. Chem.* **1** (1954); *J. Chem. Soc.* (1952) 3532.
[43] J. Koutecky and J. Paldus, *Czechoslov. Chem. Commun.* **26** (1961) 2660.
[44] Schmidt and Heinle, *Chem. Ber.* **44** (1911) 1448.
[45] M. J. S. Dewar and E. W. T. Warford *J. Chem. Soc.* (1956) 3570.
[46] S. F. Mason, *J. Chem. Soc.* (1958) 4329.
[47] See in particular: P. D. B. de la Mare, N.V. Klassen, and R. Koenigsberger, *J. Chem. Soc.* (1961) 5285.
[48] S. F. Mason, *J. Chem. Soc.* (1959) 1233.
[49] R. Daudel and A. Pullman, *Compt. Rend. Acad. Sci.* **220** (1945) 888.
[50] N. V. Swartholm, *Arkiv. Kemi, Mineral, Geol.* **15A**, No. 13 (1941).
[51] P. and R. Daudel, R. Jacques and M. Jean, **84** (1946), 489.
[52] C. A. Coulson, *Trans. Far. Soc.* **42** (1946) 106, 265.
[53] See for example: R. Daudel and A. Pullman, *J. Phys.* (1946), 109.
[54] R. Daudel, C. Sandorfy, C. Vroelant, P. Yvan, and O. Chalvet, *Bull. Soc. Chim. Fr.* **17** (1950) 66.
[55] M. Roux, *Bull. Soc. Chim.* **17** (1950) 861.
[56] F. H. Burkitt, C. A. Coulson, and H. C. Longuet-Higgins, *Trans. Far. Soc.* **47** (1951) 553.
[57] P. Yvan, *J. Chim. Phys.* **49** (1952) 457; *Compt. Rend. Acad. Sci.* **234** (1952) 2287.
[58] K. Fukui, T. Yonezawa, and C. Nagata, *J. Chem. Phys.* **26** (1957) 831.
[59] H. Baba, *Bull. Chem. Soc. Japan* **30** (1957) 147.
[60] J. Koutecky, R. Zahradnick, and J. Cizek, *Trans. Far. Soc.* **57** (1961) 169.
[61] M. Levy and M. Szwarc, *J. Am. Chem. Soc.* **77** (1955) 1949.
[62] R. Daudel and O. Chalvet, *J. Chim. Phys.* **53** (1956) 943.
[63] S. S. Sung, O. Chalvet, and R. Daudel, *J. Chim. Phys.* **57** (1960) 31.
[64] R. D. Brown, *J. Chem. Soc.* (1959), 2224, 2232.
[65] S. Nagakura and J. Tanaka, *Bull. Chem. Soc. Japan* **32** (1959) 734.
[66] S. Nagakura, *Tetrahedron* **19** (1963) 361.
[67] P. Yvan, *Compt. Rend. Acad. Sci.* **229** (1949) 622.
[68] K. Fukui, K. Morokuma, and T. Yonezawa, *Bull. Chem. Soc. Japan* **32** (1959) 1015.
[69] O. Chalvet, R. Daudel, and J. Kaufman, *J. Phys. Chem.* **68** (1964) 490.
[70] G. W. Wheland and L. Pauling, *J. Am. Chem. Soc.* **57** (1935) 2086.
[71] P. and R. Daudel, N. P. Buu-Hoi and M. Martin, *Bull. Soc. Chim.* **25** (1948) 1203.
[72] C. Sandorfy, C. Vroelant, P. Yvan, O. Chalvet, and R. Daudel, *Bull. Soc. Chim.* **17** (1950) 304.
[73] K. Fukui, T. Yonezawa, and C. Nagata, *J. Chem. Phys.* **26** (1957) 831.
[74] O. Chalvet, R. Daudel, and J. Kaufman, *J. Chem. Phys.* **68** (1964) 490.
[75] M. J. S. Dewar, T. Mole, and E. W. T. Warford, *J. Chem. Soc.* (1956) 3581.
[76] E. Heilbronner, *Non Benzenoid Aromatic Compounds*, Chapter VI, D. Ginsburg, publ., Interscience Pub., 1959.
[77] H. Arnold and K. Pahls, *Ber.* **89** (1956) 121.
[78] E. D. Bergmann, *Progress in Organic Chemistry*, Vol. III, J. W. Cook, (ed.), Ac. Press, New York, 1955, p. 81.

[79] A. S. Kende, P. T. Izzo, and W. Fulmor, *Tetrahedron Letters* (1966) 3697.
[80] B. Pullman and J. Effinger, in *Calcul des Fonctions d'Onde Moléculaires*, C.N.R.S. (ed.), 1958, p. 351.
[81] J. Fisch and H. Gilman, *Chem. Rev.* **57** (1957) 525.
[82] Personal communication.
[83] F. W. Bergstrom, *J. Org. Chem.* **2** (1937) 411.
[84] A. Kirpal and E. Reiter, *Ber.* **58** (1925) 699.
[85] J. P. Wibant and J. Nicolai, *Rec. Trav. Chim. Pays-Bas* **64** (1945) 55.
[86] S. M. McElvain and M. A. Goese, *J. Am. Chem. Soc.* **65** (1943) 2233.
[87] R. D. Brown and M. L. Hefferman, *Aust. J. Chem.* **9** (1956) 83.
[88] R. A. Barnes, *J. Am. Chem. Soc.* **81** (1959) 1935.
[89] R. D. Brown and R. D. Harcourt, *J. Chem. Soc.* (1959) 3451.
[90] M. W. Austin and J. H. Ridd, *J. Chem. Soc.* (1963), 4204.
[91] M. W. Austin, M. Brickman, J. H. Ridd, and B. V. Smith, *Chem. and Ind.* (1962), 1057.
[92] S. H. Ridd, in *Physical Methods in Heterocyclic Chemistry*, Vol. I, Ac. Press, New York, 1963, Chap. 2.
[93] R. D. Brown and R. D. Harcourt, *Tetrahedron* **8** (1960) 23.
[94] S. Kwiatkowski and B. Zurawski, *Bull. Ac. Polon. Sci.* **13** (1965) 487.
[95] R. Zahradnik and C. Parkanyi, *Czechoslov. Chem. Commun.* **30** (1965) 355.
[96] R. D. Brown and B. A. W. Coller, *Aust. J. Chem.* **12** (1959) 152.
[97] H. Hamano and H. F. Hameka, *Tetrahedron* **18** (1962) 985.
[98] M. W. Austin, J. R. Blacborov, J. H. Ridd, and B. V. Smith, *J. Chem. Soc.* (1965) 1051.
[99] L. Melander, *Arkiv för Kemi* **11** (1957) 397.
[100] R. Zahradnik and J. Koutecky, *Czechoslov. Chem. Commun.* **26** (1961) 156.
[101] See for example: P. and R. Daudel, N. P. Buu-Hoi, and M. Martin, *Bull. Soc. Chim. Fr.* **15** (1948) 1202.
[102] M. J. S. Dewar, *J. Chem. Soc. Chim.* (1949) 463.
[103] M. Brickman and J. H. Ridd, *J. Chem. Soc.* (1965) 6845.
[104] F. J. Igea Lopez-Vazquez, *Anales de la Real Sociedad Española de Fisica y Quimica* **51** B (1955) 203.
[105] B. Pullman and S. Diner, *Calcul des Fonctions d'Onde Moléculaires*, C.N.R.S., (ed.), 1958, p. 365.
[106] S. Senent and F. J. Igae, *Anales de la Real Sociedad Española de Fisica y Quimica* **53** B (1957) 403.
[107] J. Murto, Suomen Kemistilehti, B **38** (1965) 246.
[108] M. Simonetta and S. Carra, 'Nitro Compounds', *Proc. Int. Symp. of the Polish Ac. of Sciences*, Pergamon, Publ., 1964, p. 383.
[109] L. B. Kier, *J. Pharmac. Sci.* **55** (1966) 98.
[110] P. J. C. Fierens, H. Hannaert, J. Van Rysselberghe, and R. H. Martin, *Helv. Chim. Acta* **38** (1955) 2009.
[111] R. Daudel and O. Chalvet, *J. Chim. Phys.* **53** (1956) 943.
[112] R. Daudel and A. Pullman, *Compt. Rend. Acad. Sci.* **220** (1945) 888; **222** (1946) 86.
[113] R. D. Brown, *Australian J. Sci. Res.* **2**A (1949) 564; *J. Chem. Soc.* (1950) 3249; (1951) 1950.
[114] M. C. Kloetzel, *Organic Reactions*, Vol. 4, Wiley Publ., 1948, Chap. 1.
[115] H. L. Holmes, *Organic Reactions*, Vol. 4, Wiley Publ., 1948, Chapter 2.
[116] L. W. Butz and A. W. Rytinal, *Organic Reactions*, Vol. 5, Wiley Publ., 1949, Chapter 3.
[117] C. Walling, *The Chemistry of Petroleum Hydrocarbons*, Vol. 3, Reinhold Publ., 1955, Chap. 47.
[118] R. B. Woodward and T. J. Katz, *Tetrahedron* **5** (1959) 70; *Tetrahedron Letters*, No. **5** (1959) 19.
[119] M. J. S. Dewar, *Tetrahedron Letters*, No. **4** (1959) 16.
[120] R. D. Brown, *J. Chem. Soc.* (1950), 2730; *Quart. Rev.* **6** (1952) 63.
[121] R. D. Brown, *J. Chem. Soc.* (1951), 1955; (1952) 2229.
[122] O. Chalvet, R. Daudel, R. Gouarne, and M. Roux, *Compt. Rend. Acad. Sci.* **232** (1951) 2221.
[123] R. D. Brown, *J. Chem. Soc.* (1951) 3129.
[124] J. I. Fernandez-Alonso and R. Domingo, *Trans. Far. Soc.* **55** (1959) 702.
[125] M. G. Evans, *Trans. Far. Soc.* **35** (1939) 824.
[126] A. Streitwieser, *Molecular Orbital Theory for Organic Chemists*, Wiley, 1961, p. 436.
[127] E. Lehmann and W. Paasche, *Ber.* **68** (1935) 1146.

[128] See for example: R. Daudel, R. Lefebvre, and C. Moser, *Quantum Chemistry*, Interscience. 1959, p . 263; A. Streitwieser, *Molecular Orbital Theory for Organic Chemists*, Wiley, 1961, p, 438.
[129] E. Pilar, *J. Chem. Phys.* **29** (1958) 1119.
[130] K. Yang, *J. Am. Chem. Soc.* **84** (1962) 3795.
[131] T. Fueno, T. Ree, and H. Eyring, *J. Phys. Chem.* **63** (1959) 1940.
[132] A. Streitwieser and S. Suzuki, Tetrahedron **16** (1961) 153.
[133] H. Zimmerman, *Tetrahedron* **16** (1961) 169.
[134] M. Simonetta and S. Carra, *Tetrahedron* **19** (1963) 467.
[135] A. Streitwieser, Molecular Orbital Theory for Organic Chemists, Wiley, 1961, p. 445.
[136] D. A. Brown and J. R. Raju, *J. Chem. Soc.* A (1966) 40.
[137] K. Higasi, H. Baba, and A. Rembaum, *Quantum Organic Chemistry*, Interscience Publ., 1965.
[138] T. Yonezawa, K. Hayashi, C. Nagata, S. Okamura, and K. Fukui, *J. Polymer Sci.* **14** (1954) 312; *J. Polymer Sci.* **20** (1956) 537.
[139] T. Fueno, T. Tsuruta, and J. Furukawa, *J. Polymer Sci.* **40** (1959) 487.
[140] A. Pullman and R. Daudel, *Compt. Rend. Acad. Sci.* **222** (1946) 288.
[141] R. Srinivasan, *J. Am. Chem. Soc.* **85** (1963) 4045.
[142] V. A. Crawford and C. A. Coulson, *J. Chem. Soc.* (1948) 1990.
[143] J. Fernandez and R. Domingo, *Ann. Real. Soc. Esp. Fis. y Quim.* B **51** (1955) 321.
[144] E. Havinga, R. O. de Jongh, and W. Dorst, *Rec. Trav. Chim. Pays-Bas* **75** (1956) 378.
[145] H. E. Zimmermann, *Tetrahedron* **19** (1963) 397.
[146] G. Porter, 'Report to the Solvay Council', 1965.
[147] K. J. Laidler, *The Chemical Kinetics of Molecules*, Oxford Univ. Press, 1955, p. 41.
[148] E. Havinga, R. O. de Jongh, and W. Dorst, *Rec. Trav. Chim. Pays-Bas* **75** (1956) 378.
[149] R. L. Letsinger, O. B. Ramsay, and J. H. de Cain, *J.A.C.S.* **87** (1965) 2945.
[150] D. A. de Bie and E. Havinga, *Tetrahedron* **21** (1965) 2359.
[151] G. Feler, *Thèse de 3° cycle*, Paris 1967.
[152] W. G. Dauben, Report to the Solvay Council 1965.
[153] R. B. Woodward and R. Hoffmann, *J. Am. Chem. Soc.* **87** (1965) 395.
[154] H. C. Longuet-Higgins and E. W. Abrahamson, *J.A.C.S.* **87** (1965) 2045.

# INDEX OF NAMES

# INDEX OF SUBJECTS